AF800569

Springer Proceedings in Physics

Volume 408

Indexed by Scopus

The series Springer Proceedings in Physics, founded in 1984, is devoted to timely reports of state-of-the-art developments in physics and related sciences. Typically based on material presented at conferences, workshops and similar scientific meetings, volumes published in this series will constitute a comprehensive up to date source of reference on a field or subfield of relevance in contemporary physics. Proposals must include the following:

- Name, place and date of the scientific meeting
- A link to the committees (local organization, international advisors etc.)
- Scientific description of the meeting
- List of invited/plenary speakers
- An estimate of the planned proceedings book parameters (number of pages/ articles, requested number of bulk copies, submission deadline).

Please contact:

For Americas and Europe: Dr. Zachary Evenson; zachary.evenson@springer.com
For Asia, Australia and New Zealand: Dr. Loyola DSilva; loyola.dsilva@springer.com

G. Rajesh · R. Sriram · R. C. DiviaHarshaVardini
Editors

Advances in Shock Interactions

Proceedings of the 24th International Shock Interaction Symposium, SIS-2022, 17–20 October, Chennai, India

Springer

Editors
G. Rajesh
Department of Aerospace Engineering
Indian Institute of Technology Madras
Chennai, Tamil Nadu, India

R. Sriram
Department of Aerospace Engineering
Indian Institute of Technology Madras
Chennai, Tamil Nadu, India

R. C. DiviaHarshaVardini
Department of Aerospace Engineering
Indian Institute of Technology Madras
Chennai, Tamil Nadu, India

ISSN 0930-8989 ISSN 1867-4941 (electronic)
Springer Proceedings in Physics
ISBN 978-981-97-6098-5 ISBN 978-981-97-6099-2 (eBook)
https://doi.org/10.1007/978-981-97-6099-2

© The Editor(s) (if applicable) and The Author(s), under exclusive license to Springer Nature Singapore Pte Ltd. 2024

This work is subject to copyright. All rights are solely and exclusively licensed by the Publisher, whether the whole or part of the material is concerned, specifically the rights of translation, reprinting, reuse of illustrations, recitation, broadcasting, reproduction on microfilms or in any other physical way, and transmission or information storage and retrieval, electronic adaptation, computer software, or by similar or dissimilar methodology now known or hereafter developed.
The use of general descriptive names, registered names, trademarks, service marks, etc. in this publication does not imply, even in the absence of a specific statement, that such names are exempt from the relevant protective laws and regulations and therefore free for general use.
The publisher, the authors and the editors are safe to assume that the advice and information in this book are believed to be true and accurate at the date of publication. Neither the publisher nor the authors or the editors give a warranty, expressed or implied, with respect to the material contained herein or for any errors or omissions that may have been made. The publisher remains neutral with regard to jurisdictional claims in published maps and institutional affiliations.

This Springer imprint is published by the registered company Springer Nature Singapore Pte Ltd.
The registered company address is: 152 Beach Road, #21-01/04 Gateway East, Singapore 189721, Singapore

If disposing of this product, please recycle the paper.

Preface

Welcome to the proceedings of the 24th International Shock Interaction Symposium (SIS 2022), held at the prestigious Indian Institute of Technology Madras. This symposium serves as a critical forum for the presentation and discussion of cutting-edge research in the field of shock interactions, a domain that continues to challenge and inspire scientists and engineers globally.

SIS 2022 covers a wide array of topics reflecting the multidisciplinary nature of shock interaction studies, ranging from the fundamental physics of shock wave diffraction and reflection to advanced computational models and experimental techniques for understanding complex shock-shock and shock-boundary layer interactions. The papers in these proceedings represent the forefront of research and development in this field, providing new insights and innovative solutions to theoretical and practical problems.

This volume is not just a record of presented research but a testament to the hard work, dedication, and collaborative spirit of the global shock interaction community. We are deeply grateful to all the authors for their high-quality contributions and the reviewers for their rigorous and constructive feedback. Their combined efforts have been essential in maintaining the high standards of the proceedings. We also extend our heartfelt thanks to the sponsors, organizing committee, and volunteers, whose tireless efforts have been instrumental in making this symposium a success.

The research presented here not only advances our understanding of shock interactions but also paves the way for future innovations in various applications, including aerospace, defence, and energy sectors. We hope these proceedings will serve as a valuable resource for researchers and practitioners, fostering collaboration and inspiring new avenues of investigation that will contribute to the ever-expanding body of knowledge in shock interaction science.

We anticipate that the insights and thoughts presented at SIS 2022 will inspire further research and innovation. We hope you find these proceedings both stimulating and enlightening, and we encourage you to engage with the material and participate in the ongoing dialogue within the shock interaction community.

Chennai, India

G. Rajesh
Chair, SIS 2022

Contents

Contributors

R. Arun Kumar Indian Institute of Technology Jodhpur, Jodhpur, India

Elangannan Arunan Department of Inorganic and Physical Chemistry, Indian Institute of Science, Bengaluru, India

Naresh Bhatnagar Department of Mechanical Engineering, Indian Institute of Technology Delhi, New Delhi, India

Chaitanya Bhoir Department of Aerospace Engineering, Indian Institute of Science, Bengaluru, India

Kowsik V. R. Bodi Indian Institute of Technology Bombay, Mumbai, Maharashtra, India

Karthik Chilkuri Reddy Department of Aeronautical and Automobile Engineering, Manipal Institute of Technology, Manipal Academy of Higher Education, Manipal, Karnataka, India

Hemant Chouhan Department of Mechanical Engineering, Indian Institute of Technology Delhi, New Delhi, India

Subrahmanyam Duvvuri Turbulent Shear Flow Physics and Engineering Laboratory, Department of Aerospace Engineering, Indian Institute of Science, Bengaluru, India

Annie Rose Elizabeth Indian Institute of Technology Madras, Chennai, Tamil Nadu, India

Santanu Ghosh Indian Institute of Technology, Madras, Chennai, India

S. Gopalakrishnan Department of Aerospace Engineering, Indian Institute of Science, Bangalore, India

Jagadeesh Gopalan Department of Aerospace Engineering, Indian Institute of Science, Bengaluru, India

Nikhil Jacob Department of Mechanical Engineering, Government Engineering College Thrissur, Thrissur, Kerala, India;
APJ Abdul Kalam Technological University, Thiruvananthapuram, Kerala, India

G. Jagadeesh Department of Aerospace Engineering, Indian Institute of Science, Bangalore, India

Gopalan Jagadeesh Department of Aerospace Engineering, Indian Institute of Science, Bangalore, India

Sushmitha Janakiram Turbulent Shear Flow Physics and Engineering Laboratory, Department of Aerospace Engineering, Indian Institute of Science, Bengaluru, India

T. Jayachandran Indian Institute of Technology Madras, Chennai, Tamil Nadu, India

G. N. Joshi Defence Institute of Advanced Technology, Pune, Maharashtra, India

Kartikeya Kartikeya Department of Mechanical Engineering, Indian Institute of Technology Delhi, New Delhi, India

Mohd Shahid Habib Khan Indian Institute of Technology, Madras, Chennai, India

Sai Sharan Khandavilli Department of Aeronautical and Automobile Engineering, Manipal Institute of Technology, Manipal Academy of Higher Education, Manipal, Karnataka, India

A. J. Leo Department of Mechanical Engineering, Government Engineering College Thrissur, Thrissur, Kerala, India;
APJ Abdul Kalam Technological University, Thiruvananthapuram, Kerala, India

V. Lijo APJ Abdul Kalam Technological University, Thiruvananthapuram, Kerala, India;
Department of Mechanical Engineering, Government Engineering College Thrissur, Thrissur, Kerala, India

Puneet Mahajan Department of Applied Mechanics, Indian Institute of Technology Delhi, New Delhi, India

Spandan Maikap Indian Institute of Technology Jodhpur, Jodhpur, India

C. Manisankar National Aerospace Laboratories, Bengaluru, Karnataka, India

P. J. Manoj Department of Mechanical Engineering, Government Engineering College Thrissur, Thrissur, Kerala, India;
APJ Abdul Kalam Technological University, Thiruvananthapuram, Kerala, India

A. Misra Defence Institute of Advanced Technology, Pune, Maharashtra, India

Suraj Kumar Mondal Department of Aerospace Engineering, Indian Institute of Science, Bangalore, India

K. Muraleedharan Nair Department of Mechanical Engineering, Mar Baselios College of Engineering and Technology, Thiruvananthapuram, Kerala, India

T. M. Muruganandam Department of Aerospace Engineering, IIT Madras, Chennai, India

Vaisakh S. Nair Department of Mechanical Engineering, Mar Baselios College of Engineering and Technology, Thiruvananthapuram, Kerala, India

V. Nikhil Jacob Department of Mechanical Engineering, Government Engineering College Thrissur, Thrissur, Kerala, India;
APJ Abdul Kalam Technological University, Thiruvananthapuram, Kerala, India

T. Nilavarasan Defence Institute of Advanced Technology, Pune, Maharashtra, India

Mital Patel Department of Aerospace Engineering, Indian Institute of Science, Bangalore, India

Shweta Paunikar Department of Aerospace Engineering, Indian Institute of Science, Bangalore, India

Vachan D. Potluri Indian Institute of Technology Bombay, Mumbai, Maharashtra, India

Sanjay Prasad Department of Mechanical Engineering, Indian Institute of Technology Delhi, New Delhi, India

Bhalchandra P. Puranik Indian Institute of Technology Bombay, Mumbai, Maharashtra, India

Khushi Ram Department of Mechanical Engineering, Indian Institute of Technology Delhi, New Delhi, India

M. Ramakrishna Department of Aerospace Engineering, IIT Madras, Chennai, India

Shreshtha Ranjan Interdisciplinary Centre for Energy Research, Indian Institute of Science, Bangalore, India

Manish Suresh Raut Department of Aerospace Engineering, Indian Institute of Science, Bangalore, India

L. Rekha Department of Mechanical Engineering, Government Engineering College Thrissur, Kerala, India;
APJ Abdul Kalam Technological University, Kerala, India

Madhura Narayana Sabhahit Department of Aerospace Engineering, IIT Madras, Chennai, India

Rajkamal Sah Department of Aerospace Engineering, Bengaluru, India

Jatinder Pal Singh Sandhu Indian Institute of Technology, Madras, Chennai, India

Deepak Kumar Singh Department of Inorganic and Physical Chemistry, Indian Institute of Science, Bengaluru, India

S. R. Siva Prasad Kochi Department of Aerospace Engineering, IIT Madras, Chennai, India

Agnieszka M. Slowicka Institute of Fundamental Technological Research, Polish Academy of Sciences, Warszawa, Poland

R. Sriram Indian Institute of Technology Madras, Chennai, India

C. V. Sumesh Babu Department of Mechanical Engineering, Government Engineering College Thrissur, Thrissur, Kerala, India;
APJ Abdul Kalam Technological University, Thiruvananthapuram, Kerala, India

Satyam Suwas Department of Materials Engineering, Indian Institute of Science, Bangalore, India

I. V. Thara Reshma Department of Aeronautical and Automobile Engineering, Manipal Institute of Technology, Manipal Academy of Higher Education, Manipal, Karnataka, India

Siva Vayala Indian Institute of Technology Madras, Chennai, India

Harshit Verma Department of Aerospace Engineering, Indian Institute of Science, Bangalore, India

S. B. Verma National Aerospace Laboratories, Bengaluru, Karnataka, India

V. Vijayakrishnan Department of Aerospace Engineering, IIT Madras, Chennai, India

Zbigniew A. Walenta Institute of Fundamental Technological Research, Polish Academy of Sciences, Warszawa, Poland

Chapter 1
Study of Mach Reflection in Inviscid Flows

S. R. Siva Prasad Kochi and M. Ramakrishna

Abstract In this paper, we study the Mach reflection phenomenon in inviscid flows using a higher order discontinuous Galerkin method and overset grids. We use the shock capturing procedure proposed in Siva Prasad Kochi and Ramakrishna (Shock capturing using discontinuous Galerkin method and overset grids for two-dimensional Euler Equations, Journal of Computational Physics, 488, 112230, 2023) using overset grids to capture the discontinuities occurring in the supersonic flow over a wedge accurately. In this procedure, we obtain a coarse grid solution first and using the troubled cell data, we construct an overset grid which is approximately aligned to all the discontinuities. We rerun the solver with the coarse grid solution as the initial condition while using the troubled cell indicator and the limiter only on the overset grid. This allows us to capture the discontinuities accurately. Using this procedure, we have obtained the solution for Mach 3.0 and 4.0 flow over a wedge for various wedge angles and determined the detachment criterion and the Von Neumann condition accurately. We have also determined the Mach stem height for various wedge angles for these Mach numbers. We have also demonstrated the hysteresis that occurs in the transition from regular reflection to Mach reflection.

1.1 Introduction

In this paper, we study the transition between regular reflection (RR) and Mach reflection (MR) of steady shock waves in inviscid flows using discontinuous Galerkin method (DGM) along with overset grids. This has been studied quite extensively in literature experimentally [1], numerically [2], and analytically [3]. We use the shock capturing procedure proposed in [4] using overset grids and a higher order method (DGM) to capture the discontinuities occurring in the supersonic flow over a wedge accurately. In this procedure, we obtain a coarse grid solution first and using the solution and the troubled cell data, we construct an overset grid which is approximately aligned to all the discontinuities. We rerun the solver with the

S. R. S. P. Kochi (✉) · M. Ramakrishna
Department of Aerospace Engineering, IIT Madras, Chennai, India
e-mail: siva.ksr@gmail.com

© The Author(s), under exclusive license to Springer Nature Singapore Pte Ltd. 2024
G. Rajesh et al. (eds.), *Advances in Shock Interactions*, Springer Proceedings in Physics 408, https://doi.org/10.1007/978-981-97-6099-2_1

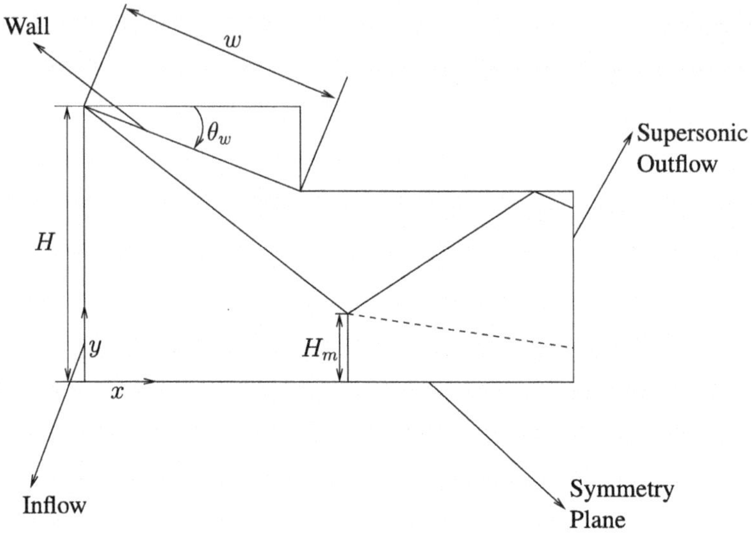

Fig. 1.1 Computational domain for Mach reflection showing the shock structure and the Mach stem height (H_m)

coarse grid solution as the initial condition while using the troubled cell indicator and the limiter only on the overset grid. This allows us to capture the discontinuities accurately. We solve the Euler equations in the computational domain shown in Fig. 1.1 using DGM and overset grids for Mach 3.0 and 4.0 flow over a wedge for different wedge angles to determine the transition between RR and MR. In this way, we determine the detachment criterion and the Von Neumann condition accurately to demonstrate the hysteresis that occurs in such a flow.

The paper is organized as follows. We describe the formulation of the discontinuous Galerkin method used for all our results in Sect. 1.2, the procedure used for shock capturing using overset grids is described in Sect. 1.3, the results are described in Sect. 1.4 and we conclude the paper in Sect. 1.5.

1.2 Description of Discontinuous Galerkin Method

Consider the Euler equations in conservative form as given by

$$\frac{\partial \mathbf{Q}}{\partial t} + \frac{\partial \mathbf{F(Q)}}{\partial x} + \frac{\partial \mathbf{G(Q)}}{\partial y} = 0 \quad \text{in the domain} \quad \Omega \tag{1.1}$$

where $\mathbf{Q} = (\rho, \rho u, \rho v, E)^T$, $\mathbf{F(Q)} = u\mathbf{Q} + (0, p, 0, pu)^T$ and $\mathbf{G(Q)} = v\mathbf{Q} + (0, 0, p, pv)^T$ with $p = (\gamma - 1)(E - \frac{1}{2}\rho(u^2 + v^2))$ and $\gamma = 1.4$. Here, ρ is the density, (u, v) is the velocity, E is the total energy and p is the pressure. We approximate the domain Ω by K non overlapping elements given by Ω_k.

We look at solving (1.1) using the discontinuous Galerkin method. We approximate the local solution in an element Ω_k, where k is the element number, as a polynomial of order N which is given by:

$$Q_h^k(r, s) = \sum_{i=0}^{N_p-1} Q_i^k \psi_i(r, s) \tag{1.2}$$

where $N_p = (N+1)(N+1)$ and r and s are the local coordinates. Here, the subscript i represents the particular degree of freedom, h represents the grid size, and the superscript k is the element number. The polynomial basis used ($\psi_i(r, s)$) is the tensor product orthonormalized Legendre polynomials of degree N. The number of degrees of freedom are given by $N_p = (N+1)(N+1)$. Now, using $\psi_j(r, s)$ as the test function, the weak form of the Eq. (1.1) is obtained as

$$\sum_{i=0}^{N_p-1} \frac{\partial Q_i^k}{\partial t} \int_{\Omega_k} \psi_i \psi_j d\Omega + \int_{\partial\Omega_k} \hat{F} \psi_j ds - \int_{\Omega_k} \mathbf{F} \cdot \nabla \psi_j d\Omega = 0 \quad j = 0, \ldots, N_p - 1 \tag{1.3}$$

where $\partial\Omega_k$ is the boundary of Ω_k, $\mathbf{F} = (\mathbf{F(Q)}, \mathbf{G(Q)})$ and $\hat{F} = \bar{F}^* \cdot \hat{n}$ where $\bar{F}^*$ is the monotone numerical flux at the interface which is calculated using an exact or approximate Riemann solver and $\hat{n}$ is the unit outward normal. This is termed to be $\mathbf{P}^N$ based discontinuous Galerkin method.
Equation (1.3) is integrated using an appropriate Gauss Legendre quadrature and is discretized in time by using the fifth order Runge-Kutta time discretization given in [5] unless otherwise specified. To control spurious oscillations which occur near discontinuities, a limiter is used with a troubled cell indicator. We have used the KXRCF troubled cell indicator [6] and the compact subcell WENO (CSWENO) limiter proposed in [7] for all our calculations.

1.3 Overset Grids and Shock Capturing

Overset grids consist of multiple grids which overlap each other as shown in Fig. 1.2. When using DGM on overset grids, there are two possible approaches to handle data communication between the grids. One is a face based communication approach developed in [8], where solutions at an overset interface are obtained from the donor element, and then the boundary condition is applied weakly by imposing a numerical flux at the flux interpolation points. The other is an element based communication approach developed in [9], where the internal degrees of freedom of cells near the overset interface are obtained from the donor element. We use the new element based communication approach developed in [10] for the data communication between the grids.

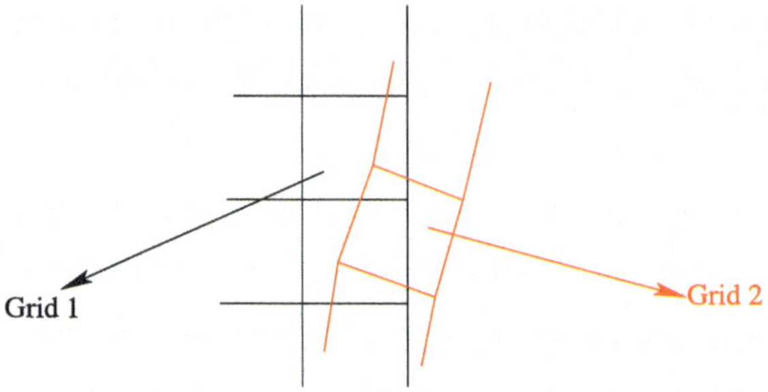

Fig. 1.2 Two overlapping grids (Grid 1 in black and Grid 2 in Red)

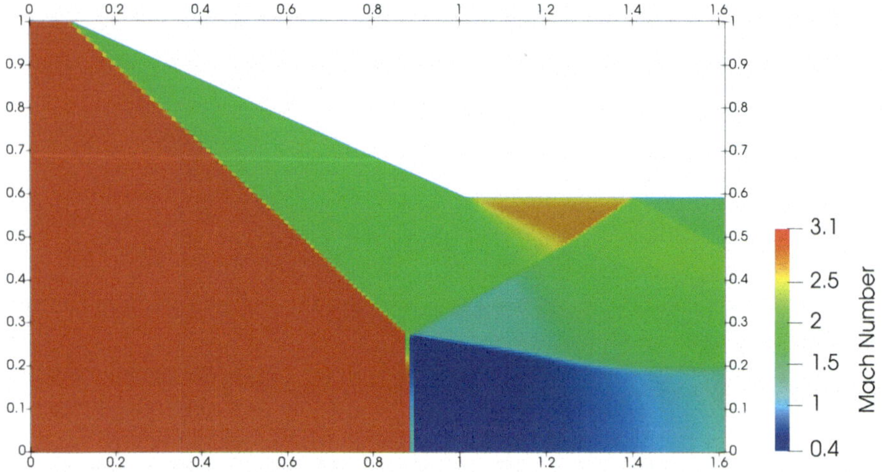

Fig. 1.3 Coarse grid solution for Mach 3.0 flow over a 24° wedge using $\mathbf{P}^1$ based discontinuous Galerkin method

For capturing the shocks accurately, we use the procedure developed in [4] using overset grids. A brief explanation of the procedure is given below:

Step 1: Run the solver on a coarse grid with a given troubled cell indicator and limiter to steady state and obtain the solution. As an example, we show the coarse grid solution obtained for Mach 3.0 flow over a 24° wedge in Fig. 1.3

Step 2: Look at the troubled cells to locate the discontinuities (shocks) that occur in the solution. The troubled cell profile obtained for the Mach 3.0 flow over a 24° wedge using the KXRCF troubled cell indicator [6] is shown in Fig. 1.4. From this figure, we can see that the troubled cells give us a good idea of the location of the shocks, the contact discontinuity and the initial expansion that forms near the wedge.

Step 3: Construct an overset grid conforming to the computational domain which is refined in a direction perpendicular to the discontinuities such that they are approximately parallel to a grid line. This overset grid also encompasses all the troubled

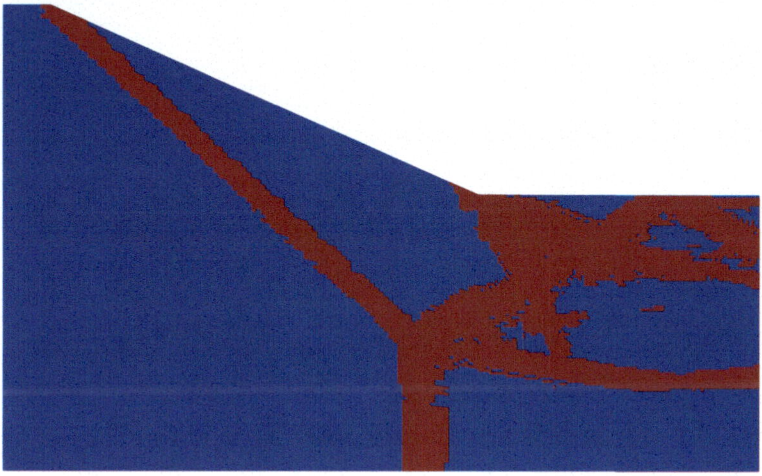

Fig. 1.4 Troubled cells obtained using the KXRCF troubled cell indicator [6] for Mach 3.0 flow over a 24° wedge

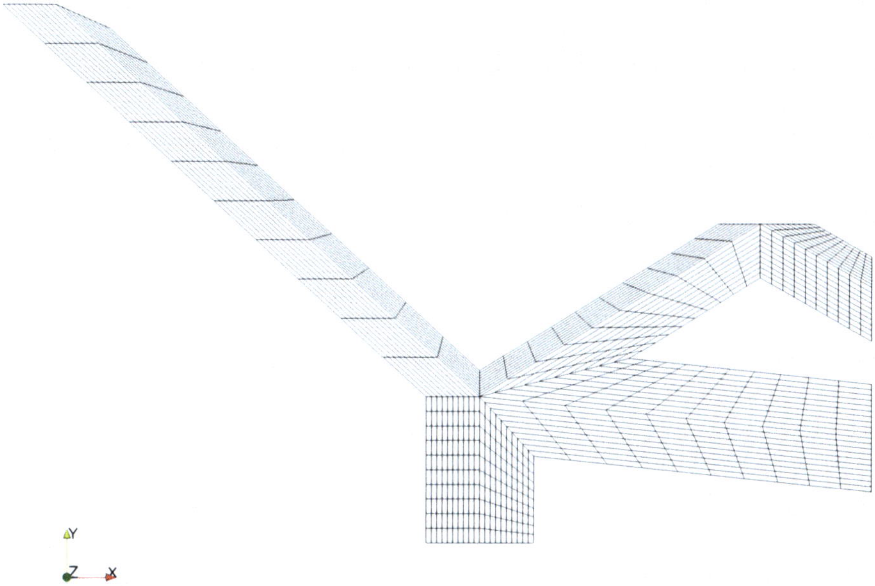

Fig. 1.5 Example overset grid for Mach 3.0 flow over a 24° wedge constructed such that the discontinuities are approximately parallel to a grid line

cells. An example overset grid constructed in such a fashion for the Mach 3.0 flow over a 24° wedge is shown in Fig. 1.5.

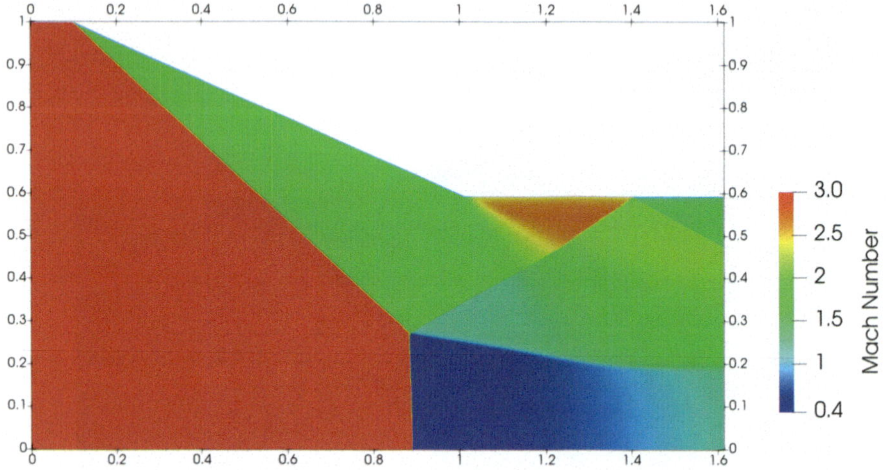

Fig. 1.6 Final solution obtained for Mach 3.0 flow over a 24° wedge using an overset grid and $\mathbf{P}^4$ based discontinuous Galerkin method with discontinuities captured accurately

Step 4: Using this overset grid, we rerun the solver with the coarse grid solution as the initial condition. While running the solver, we use the troubled cell indicator and the limiter only on the overset grid. We also use a high resolution numerical flux on the overset grid to capture the shock accurately. We have used the SLAU2 [11] numerical flux in the overset grid and the less expensive Lax-Friedrichs flux elsewhere. Using this procedure, we obtain a more accurate solution with the discontinuities approximately aligned to a grid line. The final solution obtained in this fashion for Mach 3.0 flow over a 24° wedge is shown in Fig. 1.6.

Regarding the computational cost of the scheme proposed for the current problem, we note that the troubled cell data shown in Fig. 1.4 are obtained from a coarse grid of 20000 elements using $\mathbf{P}^1$ based DGM after converging to a residue of about 1×10^{-6} which happens in about 8000 iterations. This coarse grid solution is not a fully converged solution but is good enough to be used as an initial condition for the overset grid higher order ($\mathbf{P}^4$ based DGM) solution. The time taken per iteration per degree of freedom using $\mathbf{P}^1$ based DGM for this problem is about 1.084×10^{-6}s. This tells us that the coarse grid troubled cell data is obtained in about 693.76s which is a little less than 12 min. Most of the work involved is in constructing the overset grid which is done manually. After the overset grid is constructed and the solver is run on the new grid with the coarse grid solution as the initial condition, the residue converges to 1×10^{-16} in about 32000 iterations. For $\mathbf{P}^4$ based DGM, the time taken per iteration per degree of freedom is about 1.345×10^{-6}s for this problem. This tells us that the final solution (which is fifth order accurate) is obtained in about 21520s which is about six hours. All these calculations are done on a 3.60 GHz, Intel(R) Core(TM) i7-7700 CPU with a single thread. The calculations for remaining angles are quite similar. We also note that we have a parallelised code and we obtain the solution much faster based on the number of threads used.

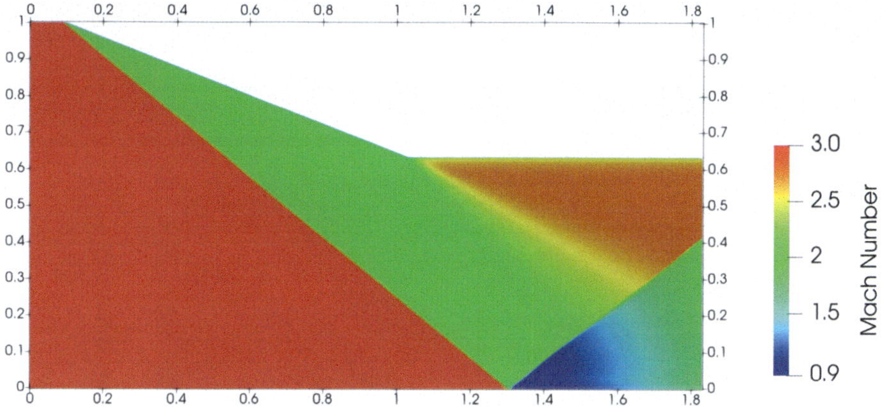

Fig. 1.7 Final solution obtained for Mach 3.0 flow over a 21.45° wedge with an overset grid and $\mathbf{P}^4$ based discontinuous Galerkin method with discontinuities captured accurately using an impulsive start

1.4 Results

(1) Mach number 3.0: We solve the two-dimensional Euler equations given by (1.1) using the discontinuous Galerkin method for various wedge angles between $\theta_w = 19.5°$ and $\theta_w = 24°$ near the transition criterion (transition from Regular reflection to Mach reflection) for Mach number 3.0 and $w/H = 1.0$ in the computational domain shown in Fig. 1.1. We consider two cases to demonstrate the hysteresis phenomenon. We solve the equations using an impulsive start as the initial condition for the first case, and the converged solution for $\theta_w = 24°$ as the second case. The first case is a numerical model to obtain the detachment criterion and the second case demonstrates the Von Neumann condition. Using the procedure outlined in Sect. 1.3, we obtain a very accurate solution for each of the wedge angles and obtain the transition criterion clearly. Note that the initial conditions mentioned above are used to obtain the coarse grid solution first and the coarse grid solution is used as an initial condition to get the fine grid solution using $\mathbf{P}^4$ based DGM. To show that the transition criterion has been captured accurately, we show the solution obtained for Mach 3.0 flow over a wedge with wedge angles $\theta_w = 21.45°$ and $\theta_w = 21.46°$ with the case 1 initial conditions in Figs. 1.7 and 1.8 respectively.

We also show the table containing the Mach stem height obtained using this procedure for various wedge angles for both cases of initial conditions in Table 1.1. From Table 1.1, we can see that the transition criterion occurs between wedge angles 21.45° and 21.46° and the Von Neumann transition criterion occurs between wedge angles 19.6° and 19.7°. From the three-shock theory, the transition criterion is at $\theta_w = 21.456°$ and the Von Neumann condition occurs at $\theta_w = 19.656°$. This closely agrees with what we have obtained and validates our procedure.

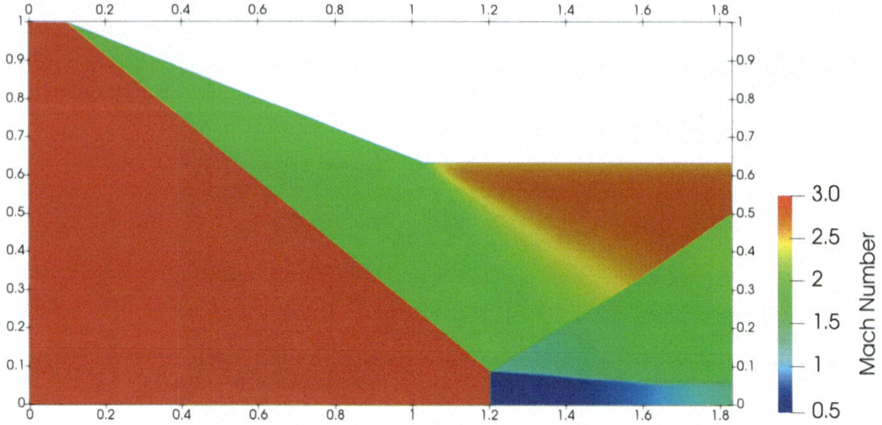

Fig. 1.8 Final solution obtained for Mach 3.0 flow over a 21.46° wedge with an overset grid and $\mathbf{P}^4$ based discontinuous Galerkin method with discontinuities captured accurately using an impulsive start

Table 1.1 Non-dimensionalised height of the Mach stem (H_m/H) as a function of wedge angle (θ_w) for M=3.0 and $w/H = 1.0$ using two sets of initial conditions:- Case 1: Impulsive start, Case 2: Converged solution for $\theta_w = 24°$

θ_w	H_m/H for case 1 (impulsive start)	H_m/H for case 2 (converged solution for $\theta_w = 24°$)
24°	0.274	–
23.5°	0.233	0.233
23°	0.188	0.188
22.5°	0.153	0.153
22°	0.122	0.122
21.46°	0.078	0.078
21.45°	RR	0.066
21°	RR	0.041
20.5°	RR	0.028
20°	RR	0.009
19.7°	RR	0.002
19.6°	RR	RR
19.5°	RR	RR

Table 1.2 Non-dimensionalised height of the Mach stem (H_m/H) as a function of wedge angle (θ_w) for M=4.0 and $w/H = 1.0$ using two sets of initial conditions:- Case 1: Impulsive start, Case 2: Converged solution for $\theta_w = 27°$

θ_w	H_m/H for case 1 (impulsive start)	H_m/H for case 2 (converged solution for $\theta_w = 27°$)
27°	0.344	-
26.5°	0.313	0.313
26°	0.285	0.285
25.7°	0.245	0.245
25.6°	RR	0.224
25.5°	RR	0.203
25°	RR	0.183
24.5°	RR	0.144
24°	RR	0.115
23°	RR	0.093
22°	RR	0.045
21°	RR	0.005
20.9°	RR	0.002
20.8°	RR	RR
20.5°	RR	RR

(2) Mach number 4.0: We repeat the same procedure for various wedge angles between $\theta_w = 20.5°$ and $\theta_w = 27°$ near the transition criterion (transition from Regular reflection to Mach reflection) for Mach number 4.0 and $w/H = 1.0$ in the computational domain shown in Fig. 1.1. Again, we consider two cases to demonstrate the hysteresis phenomenon. We solve the equations using an impulsive start as the initial condition for the first case, and the converged solution for $\theta_w = 27°$ as the second case. We show the table containing the Mach stem height obtained using this procedure for various wedge angles for both cases of initial conditions in Table 1.2. From Table 1.2, we can see that the transition criterion occurs between wedge angles 25.6° and 25.7° and the Von Neumann transition criterion occurs between wedge angles 20.9° and 20.8°. From the three-shock theory, the transition criterion for Mach number 4.0 is at $\theta_w = 25.61°$ and the Von Neumann condition occurs at $\theta_w = 20.86°$. This closely agrees with what we have obtained and validates our procedure.

1.5 Conclusion

We have demonstrated the transition between regular reflection (RR) and Mach reflection (MR) of steady shock waves in inviscid flows using discontinuous Galerkin method (DGM) along with overset grids using an accurate shock capturing procedure [4]. We have identified the detachment criterion and the Von Neumann condition accurately and demonstrated the hysteresis which occurs in the transition. We have also calculated the Mach stem height for various wedge angles for Mach numbers 3.0 and 4.0. As future work, this shock capturing procedure can be used to further study the different flow phenomena that occur when the inflow Mach number is small (< 2.0).

References

1. A. Chpoun, D. Passerel, H. Li, G. Ben-Dor, Reconsideration of the state-of-the-art of oblique shock wave reflection in steady flows. Part I: Experimental investigation. J. Fluid Mech. **301**, 19–35 (1995)
2. J. Vuillon, D. Zeitoun, G. Ben-Dor, Reconsideration of the state-of-the-art of oblique shock wave reflection in steady flows. Part 2: numerical investigation. J. Fluid Mech. **301**, 37–50 (1995)
3. G. Ben-Dor, *Shock wave reflection phenomena* (Springer, New York, 2007)
4. S.R. Siva Prasad Kochi, M. Ramakrishna, Shock capturing using discontinuous Galerkin method and overset grids for two-dimensional Euler equations. J. Comput. Phys. **488**, 112230 (2023)
5. J.C. Butcher, Numerical methods for ordinary differential equations. 3rd edn., John Wiley and Sons, (2016)
6. L. Krivodonova, J. Xin, J.-F. Remacle, N. Chevaugeon, J. Flaherty, Shock detection and limiting with discontinuous Galerkin methods for hyperbolic conservation laws. Appl. Numer. Math. **48**, 323–338 (2004)
7. S.R. Siva Prasad Kochi, M. Ramakrishna, A compact subcell WENO limiting strategy using immediate neighbours for Runge-Kutta discontinuous Galerkin methods. Int. J. Comput. Math. **98**(3), 608–626 (2021)
8. M.C. Galbraith, J.A. Benek, P.D. Orkwis, M.G. Turner, A discontinuous Galerkin Chimera scheme. Comput. Fluids **98**, 27–53 (2014)
9. C.R. Nastase, D.J. Mavriplis, J. Sitaraman, An overset unstructured mesh discontinuous galerkin approach for aerodynamic problems. AIAA 2011-195 (2011)
10. S.R. Siva Prasad Kochi, M. Ramakrishna, A discontinuous Galerkin overset scheme using WENO reconstruction and subcells for twodimensional problems. J. Sci. Comput. **93**, 35 (2022)
11. K. Kitamura, E. Shima, Towards shock-stable and accurate hypersonic heating computations: A new pressure flux for AUSM-family schemes. J. Comput. Phys. **245**, 62–83 (2013)

Chapter 2
An Experimental Investigation of Shock Behavior of Glass and Basalt Fiber Composite Laminates in a Shock Tube

Khushi Ram, Kartikeya Kartikeya, Hemant Chouhan, Sanjay Prasad, Puneet Mahajan, and Naresh Bhatnagar

Abstract Improvised explosive device (IED) blasts causes thousands of deaths every year along with damage to infrastructure. In this study, the deformation behavior of popular blast mitigating materials, i.e., glass fiber reinforced polypropylene (G/PP) and basalt fiber reinforced polypropylene (B/PP) composites, is explored. G/PP and B/PP laminates were fabricated using compression molding and subjected to shock loading using an in-house developed shock tube. The peak-over-pressure (POP), deformation of laminates and strain generated during shock loading were recorded and analyzed. G/PP laminates experienced more deformation and failed by fiber breakage and fiber pull-out. B/PP laminates exhibited enhanced shock wave mitigation properties at the same pressure and revealed less fiber damage. This study gives an idea about the blast mitigation behavior of composite laminates and provides the direction for future work.

2.1 Introduction

In recent years, understanding of dynamic behavior of composites has been gain attention due to their inherent properties such as high strength-to-weight ratio, good stealth materials, high stiffness, and low-density properties, etc. [1]. Due to such superior properties composite materials demand has increased in defense sector to save human lives against shock loading. Most of the studies available in the literature characterize the dynamic behavior of homogenous materials. The advent of

K. Ram · K. Kartikeya · H. Chouhan · S. Prasad · N. Bhatnagar (✉)
Department of Mechanical Engineering, Indian Institute of Technology Delhi, New Delhi 110016, India
e-mail: nareshb@mech.iitd.ac.in

P. Mahajan
Department of Applied Mechanics, Indian Institute of Technology Delhi, New Delhi 110016, India

© The Author(s), under exclusive license to Springer Nature Singapore Pte Ltd. 2024
G. Rajesh et al. (eds.), *Advances in Shock Interactions*, Springer Proceedings in Physics 408, https://doi.org/10.1007/978-981-97-6099-2_2

composite materials has channeled interest in understanding the damage behavior of these materials against shock loading also [2].

Comtois et al. [3] studied the effect of standoff distance and charge mass on different thicknesses of carbon and glass fiber composites. Results showed that carbon fibers deformed less as compared to glass fiber for the same mass of charge and standoff distance. Tekalur et al. [4] analyzed the quasi-static and high strain rate properties of E-glass and carbon fiber reinforced vinyl ester composites. Authors concluded that carbon fiber composites were stiffer than E-glass fiber composites since carbon fiber composites did not exhibit any damage up to a shock pressure of 0.6 MPa. Yahya et al. [5] manufactured carbon-fiber reinforced epoxy composites of 2.2, 4.5 and 8.4 mm thickness which was tested against plastic explosive (PE4) from 1 to 12 gm. They found that due to blast buckling, fracture and delimitation of fibers had taken place. Kumar et al. [6] examined the effect of curvature on carbon fiber composites under shock loading. They fabricated three panels one having an infinite radius of curvature second one with 305 mm radius of curvature and the third one having 112 mm radius of curvature using unidirectional carbon fiber via autoclave technique. Shock tube study showed that the threshold value of pressure for panel A is 3,65 MPa, for B is 3 MPa and for C its 7.78 MPa it means up to these pressure values panel does not fail but beyond this value, panels are failed catastrophically. They also recorded blast impulse for each panel which is 14, 6.7 and 29 kPa-s for A, B and C panel. Digital image correlation (DIC) results showed that panel B deformed more as compared to A and C and which is verified by threshold pressure and by blast impulse. They finally concluded that there is an optimum radius of curvature at which panels performed best against the shock loading.

Basturk et al. [7] developed an analytical model to study the behavior of basalt fiber laminates under blast loading. They drive the equation of motion using the virtual work principle then they solve the problem using the Newmark method and compared the deflection results with finite element results obtained using Ansys software. They found basalt fiber laminates showed less center point displacement as compared to glass fiber laminates and the thermotical results match with simulation results Huang et al. [8] studied the effect of shock loading on unidirectional and cross-ply flax fiber reinforced polymer composites. They also compared the effect of polymer matrix on blast response of composite. Results showed flax/polypropylene composites had better performance against shock loading as compared to flax epoxy composite laminates. In polypropylene laminates, delamination occurred which absorbed energy before failure. Secondly, they observed cross-ply laminates behaved better as compared to unidirectional laminates because they are stiffer and withstand up to high peak overpressure loadings.

Karahan and Karahan [9] investigated three types of boot soles Type1 Monolithic steel, Type2 Aluminium honeycomb and Type3. Results indicated that Aluminium honeycomb absorbed more energy as compared to others against 40 gm PE4 explosive. Dear et al. [10] analyzed sandwich panels using three different polymeric foams and with CFRP and GFRP sheets Results indicate that SAN was more resistant to the momentum. Gargano et al. [11] studied the performance of glass and carbon fiber composites against air blast with polyester and vinyl ester as a matrix material. They tested laminates against far-field and near-field conditions. Results indicate center point deflection for carbon fiber reinforced polyester composites was more than for glass fiber reinforced polyester composites. Similar results were obtained with vinyl ester as a matrix material. The study concludes that glass fiber exhibited superior properties against blast loading as compared to carbon fiber composites similarly vinyl ester laminates showed less fiber matrix delamination when viewed under ultrasonic scanning than polyester matrix laminates. Danilov et al. [12] analyzed experimentally and numerically the effect of blast loading on duraluminium (D16) plates having 1.5 mm thickness. From literature it is evident that GFRP, CFRP, BFRP composites have been studied using thermosetting resin as a matrix material. Insufficient experimental studies are available in the open literature focusing on the shock behavior of glass and basalt fiber reinforced polypropylene composites. Accordingly, the present study is focused on the shock loading of glass and basalt fiber reinforced polypropylene composites, fabricated by compression molding technique.

2.2 Materials and Methods

2.2.1 Materials

Glass fiber and basalt fiber were chosen as the reinforcement material and polypropylene was used as a matrix material for the fabrication of composite laminates. Glass fiber and basalt fiber are available in the form of bi-directional fabric. Polypropylene is available in the form of granules which are converted into the form of thin films using a single screw extruder.

2.2.2 Fabrication of Composite Laminates

Composite laminates of glass and basalt fiber are fabricated using a vacuum compression molding machine. Fibers and polypropylene sheets are cut in 200 $\times$ 200 mm dimensions one layer of fiber and one layer of a matrix then to desired thickness is molded. The compression molding was done at 100 bar pressure and at 200 °C temperature for both the composite system and vacuum is maintained throughout the molding cycle. Using a micrometer thickness of each panel is measured at various points and it is found that 0.91 $\pm$ 0.1 mm and 1.12 $\pm$ 0.1 mm for G/PP and B/PP composites, respectively.

2.2.3 *In-house Designed and Developed Shock Tube*

Conventionally, there are two different ways to generate the shock waves for understanding the dynamic behavior of any material; one is the detonation of an explosive and the second is a shock tube. Detonation of explosives is an easy method for the generation of shock waves. However, in this method, three-dimensional waves are formed which are difficult to capture and analyze. Along with this during the explosion visualization of the wave propagation and repeatability of the test makes this method difficult. The shock tube is the one another to be used for a study of a dynamic response of any materials inside a lab. A shock tube consists of a long cylindrical tube that is further divided into two sections, one is the high-pressure section known as the driver section and the other is the low-pressure section known as the driven section. Driver and driven section are separated by a diaphragm. The shock tube used for this study is shown in Fig. 2.1. The driver section is filled with pressurized air and when the pressure difference between the driver and driven section reaches pre-determined values, the diaphragm ruptures. The driven section is at an ambient atmospheric pressure, the sudden rupture of the diaphragm generates a number of compression waves in the driven section and then these waves meet at a certain distance and forms a shock wave that hits the target material at the long end of the driven section. The distance at which shock formation takes place is known as shock formation distance, and after this distance shock wave moves smoothly is one dimension.

The shock tube is instrumented with four pressure sensors to measure the shock wave pressure and velocity. One pressure sensor is mounted at the driver section to measure the actual value of pressure at which the diaphragm rupture and the rest three are mounted at the driven section to measure shock wave pressure and its velocity. The shock pressure value obtained depends upon the thickness of the diaphragm. A

Fig. 2.1 In house designed and developed shock tube

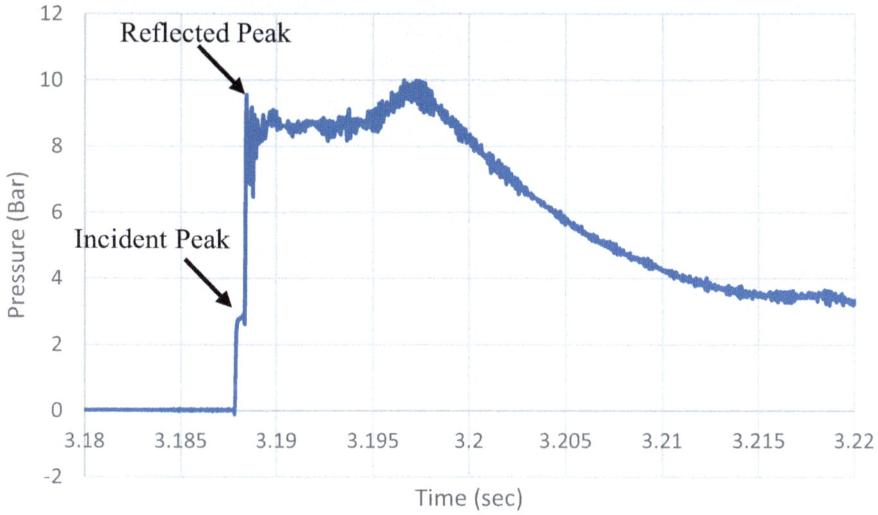

Fig. 2.2 Shock wave pressure time history

shock wave pressure time history obtained from the shock tube is plotted in Fig. 2.2. The first peak of the shock pressure is the incident shock wave to which the specimen is subjected under the given test conditions. The second pressure peak is known as the reflected shock wave which comes from the target material. This reflected shock wave varies from material to material and the specimen mounting conditions.

The laminates are mounted at the end of the driven end using a circular disc and a C-clamp. A 100 mm diameter area of specimen observes the shock loading. Strain gauges are mounted on each specimen in the horizontal direction (X1) and in the vertical direction (X2) to measure the strain when the shock wave impacts the specimen. Figure 2.3 shows a sample of the test specimen used for this study.

2.3 Results and Discussion

The shock loading behavior of 6-layer glass and basalt fiber reinforced polypropylene laminate were studied and analyzed under increasing shock wave pressure by varying the driver section pressure in the shock tube. Table 2.1 shows the shock loading pressure and the center point deflection observed by the laminates. The driver pressure varies from 7.5 bar to 27 bar, corresponds to which center point deflection is measured using height and dial gauges with the help of a surface plate. After the shock loading, each laminate is studied visually, and observed that at low shock pressure, laminates deform at the center region only. Similarly, at high pressure, the laminates deformed at the center as well as at boundary. The maximum deformation occurred at the center which reduces when one moves toward the boundary.

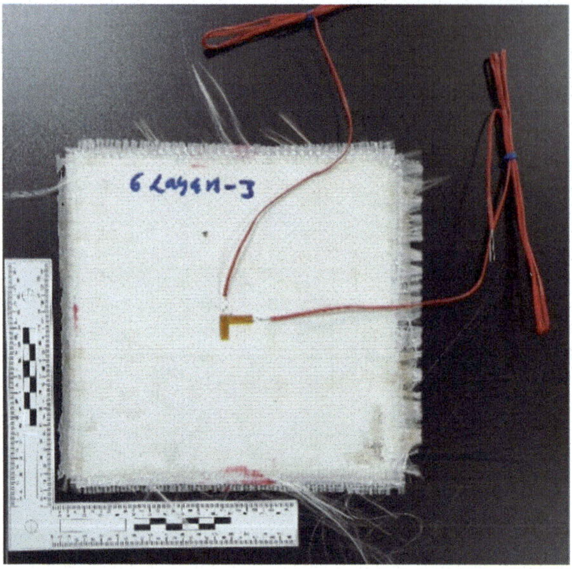

Fig. 2.3 Specimen used in the study

Table 2.1 Shock impact results for glass and basalt fiber laminates

G/PP-6 layer			B/PP-6 layer	
S.No.	Driver pressure (Bar)	Center point deflection (mm)	Driver pressure (Bar)	Center point deflection (mm)
1	7.5	9.9	7.2	7.2
2	15.3	12.3	15.1	10.6
3	22.2	Specimen ruptured	22.5	13.3
4			27.8	Specimen ruptured

Figure 2.4a shows the initial mounting of the specimen which is fully clamped with the help of disc and C-clamps. The two strain gauges in X1 and X2 directions are pasted on the specimen and their data is recorded with the help of a strain conditioning unit. Figure 2.4b shows the deformation of G/PP laminate at 15.3 bar pressure and it is observed that the specimen is deformed and maximum deformation occurred at the center and it is clearly visible that the fibers are elongated from the boundary also. From Figure 2.4c it is observed that the deformation at boundary lines is not uniform fibers, stretch from all sides but in one region, they stretches more and on the other side they stretched less, this may happen due to the boundary conditions of the specimen when it is mounted.

Figure 2.5a shows the failed specimen of G/PP laminate at 22.2 bar. Its seen that fibers are failed and delamination of fibers takes place due to the shock loading. In some region, shearing of fibers take place and at one point it is observed that a laminate

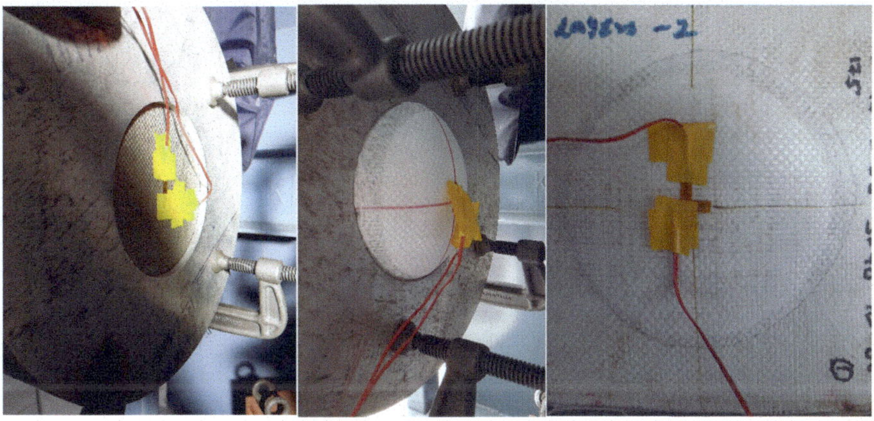

Fig. 2.4 Test Specimen **a** Initial mounting of the specimen **b** G/PP specimen after test **c** Damaged at clamped boundary edges

is delaminated and fiber failure occurred only in half region and at some region, laminate shows fiber fracture at the clamping boundary only and not at the edges of laminates. Similarly in B/PP laminate fiber fracture and delamination occurred at 27.8 bar but the laminate shows less damage as compared to G/PP. There is a threshold pressure value for each G/PP and B/PP laminates up to which they showed permanent deformation and fiber fracture at clamped boundaries but beyond this threshold pressure the laminates do not sustain the shock pressure and completely fail via fiber fracture, delamination, and shearing.

Figure 2.6 shows the strain gauge response of the laminate used in this study. One major observation with strain gauges is that at low-pressure, data is recorded easily but at higher pressure data recording is difficult and when laminates fail no data is recorded. Strain gauge data show less deformation in B/PP as compared to G/PP laminates.

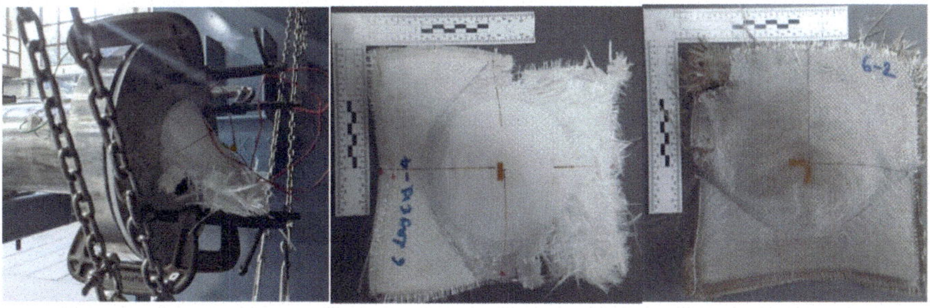

Fig. 2.5 Failed specimen **a** G/PP at 22.5 bar **b** G/PP failed specimen **c** B/PP deformed specimen

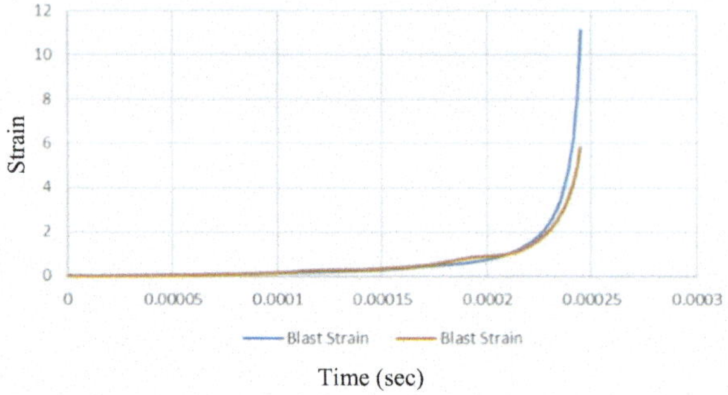

Fig. 2.6 Strain gauge response of the laminate used in this study

2.4 Conclusion

Under shock loading impact failure behavior of composites is studied and following conclusions are carried out:

1. For shock loading a threshold pressure value up to which laminates sustain the impact energy and beyond which it fails. For G/PP threshold pressure is 15.3 bar and for B/PP it is 22.5 bar.
2. Under shock loading shear mode of failure occurred at high pressure and in some laminates delamination of fibers also observed.
3. It is recommended that for strain measurement under shock impact strain gauges are not appropriate thus digital image correlation method is better which will be used in further studies.
4. B/PP laminates can be used for design of blast-resistant material system due to its capability to absorb shock energy.

References

1. A.P. Mouritz, Advances in understanding the response of fibre-based polymer composites to shock waves and explosive blasts. Compos. Part A Appl. Sci. Manuf. **125**, 105502 (2019). https://doi.org/10.1016/j.compositesa.2019.105502
2. P. Wanchoo, H. Matos, C.E. Rousseau, A. Shukla, Investigations on air and underwater blast mitigation in polymeric composite structures—a review. Compos. Struct. **263**, 113530 (2021). https://doi.org/10.1016/j.compstruct.2020.113530
3. J.L.R. Comtois, M.R. Edwards, M.C. Oakes, The effect of explosives on polymer matrix composite laminates. Compos. Part A Appl. Sci. Manuf. **30**, 181–190 (1999). https://doi.org/10.1016/S1359-835X(98)00172-9
4. S.A. Tekalur, K. Shivakumar, A. Shukla, Mechanical behavior and damage evolution in E-glass vinyl ester and carbon composites subjected to static and blast loads. Compos. Part B Eng. **39**, 57–65 (2008). https://doi.org/10.1016/j.compositesb.2007.02.020

5. M.Y. Yahya, W.J. Cantwell, G.S. Langdon, G.N. Nurick, The blast resistance of a woven carbon fiber-reinforced epoxy composite. J. Compos. Mater. **45**, 789–801 (2011). https://doi.org/10.1177/0021998310376103
6. P. Kumar, D.S. Stargel, A. Shukla, Effect of plate curvature on blast response of carbon composite panels. Compos. Struct. **99**, 19–30 (2013). https://doi.org/10.1016/j.compstruct.2012.11.036
7. S. Baştürk, H. Uyanik, Z. Kazanci, An analytical model for predicting the deflection of laminated basalt composite plates under dynamic loads. Compos. Struct. **116**, 273–285 (2014). https://doi.org/10.1016/j.compstruct.2014.05.018
8. K. Huang, A.V. Rammohan, U. Kureemun, W.S. Teo, L.Q.N. Tran, H.P. Lee, Shock wave impact behavior of flax fiber reinforced polymer composites. Compos. Part B Eng. **102**, 78–85 (2016). https://doi.org/10.1016/j.compositesb.2016.07.014
9. M. Karahan, E.A. Karahan, Development of an innovative sandwich composite material for protection of lower limb against landmine explosion: mechanical leg test results. Text. Res. J. **87**, 15–30 (2017). https://doi.org/10.1177/0040517515624880
10. J.P. Dear, E. Rolfe, M. Kelly, H. Arora, P.A. Hooper, Blast performance of composite sandwich structures. Procedia Eng **173**, 471–478 (2017). https://doi.org/10.1016/j.proeng.2016.12.065
11. A. Gargano, K. Pingkarawat, M. Blacklock, V. Pickerd, A.P. Mouritz, Comparative assessment of the explosive blast performance of carbon and glass fibre-polymer composites used in naval ship structures. Compos. Struct. **171**, 306–316 (2017). https://doi.org/10.1016/j.compstruct.2017.03.041
12. M.N. Danilov, V.V. Adishchev, N.N. Fedorova, Study of thin-walled structure behavior under shock-wave loading, AIP Conf. Proc. **1893** (2017). https://doi.org/10.1063/1.5007595

Chapter 3
Numerical Investigation of the Von Neumann Paradox for Weak Shocks Using a Higher Order Discontinuous Galerkin Method and Overset Grids

S. R. Siva Prasad Kochi and M. Ramakrishna

Abstract In this paper, we investigate the Guderley Mach reflection that occurs when a weak shock reflects off a plane wedge numerically using a higher order discontinuous Galerkin method and overset grids. Guderley Reflection (existence of a single supersonic patch near the triple point) and Guderley Mach reflection (existence of multiple supersonic patches near the triple point) are the suggested solutions for the resolution of the von Neumann paradox in the weak shock reflection. We confirm the existence of multiple supersonic patches near the triple point for three different configurations. We observe that refining the grid near the triple point increases the number of supersonic patches that occur and Guderley Mach reflection occurs in each case. We have also observed that the number of supersonic patches further increase if we add viscous fluxes near the triple point. These numerical solutions lend credence to the observation reported in Tesdall et al. (JSC 64:721–744, 2015) that the Guderley reflection is not a distinct reflection geometry and it is only an under resolved Guderley Mach reflection.

3.1 Introduction

There are three basic patterns of reflection for the reflection of weak to moderately strong shocks off a plane wedge: regular reflection, Mach reflection, and von Neumann reflection. The classical theory which describes regular and Mach reflection is called two-shock and three-shock theory respectively. For regular and Mach reflection, agreement between the theory and experiment is quite good. However, it has been observed experimentally [1] that the irregular reflection of a weak shock off a ramp shows a reflection pattern that resembles a Mach reflection with a triple point. But, according to the classical shock theory, for weak shocks, three shock solutions do not exist. This discrepancy in theory and experiment has come to be known as the von Neumann paradox.

S. R. S. P. Kochi (✉) · M. Ramakrishna
Department of Aerospace Engineering, IIT Madras, Chennai, India
e-mail: siva.ksr@gmail.com

© The Author(s), under exclusive license to Springer Nature Singapore Pte Ltd. 2024

G. Rajesh et al. (eds.), *Advances in Shock Interactions*, Springer Proceedings in Physics 408, https://doi.org/10.1007/978-981-97-6099-2_3

To resolve this paradox, Guderley [2] suggested that there is a supersonic patch behind the triple point in a Mach reflection with a weak shock, and that an expansion fan is generated at the triple point. This has come to be known as Guderley reflection (GR) pattern. Recently, numerical evidence for this existence of supersonic patch and expansion fan has been found by Vasil'ev and Kraiko [3]. It has also been suggested by Tesdall and Hunter [4] that multiple supersonic patches exist near the triple point. This multiple supersonic patch pattern is known as Guderley Mach reflection (GMR) in literature. Experiments conducted by Skews et al. [5] and recent numerical simulations conducted by Tesdall et al. [6] support this multiple patch assumption.

In this paper, we investigate this phenomenon numerically by solving 2D Euler equations using a higher order discontinuous Galerkin method (DGM) and a very fine overset grid near the triple point to confirm the existence of multiple supersonic patches. We also investigate the effect of physical viscosity on these patches by solving the Navier-Stokes equations only near the triple point using an overset grid. We solve the 2D Euler equations (or 2D Navier-Stokes equations based on the problem) in the computational domain shown in Fig. 3.1 where a moving shock of Mach number M is located at $x = 0.4$ initially. The left and right states of the flow are prescribed based on the Mach number M. We consider three different configurations with three different Mach numbers and calculate the solutions corresponding to three cases for each Mach number: 1) grid of size $h = 1/800$; 2) $h = 1/800$ grid and an overset grid of $h = 1/1600$ near triple point; and 3) $h = 1/800$ grid and an overset grid of $h = 1/1600$ near triple point with viscous fluxes applied only on the overset grid; using a fifth order accurate discontinuous Galerkin method. We draw the sonic line near the triple point for each of the cases considered and confirm the existence of supersonic patches. We also conclude that the number of supersonic patches increase if we refine the grid and also if we apply the viscous fluxes.

The paper is organized as follows. We describe the formulation of the discontinuous Galerkin method used for all our results in Sect. 3.2, procedure for using overset grids is described in Sect. 3.3, the results are described in Sect. 3.4 and we conclude the paper in Sect. 3.5.

3.2 Formulation of Discontinuous Galerkin Method

Consider the Navier-Stokes equations as given by

$$\frac{\partial \mathbf{Q}}{\partial t} + \frac{\partial}{\partial x}(\mathbf{F}(\mathbf{Q}) - \mathbf{F}_v(\mathbf{Q})) + \frac{\partial}{\partial y}(\mathbf{G}(\mathbf{Q}) - \mathbf{G}_v(\mathbf{Q})) = 0 \quad \text{in the domain} \quad \Omega \quad (3.1)$$

where $\mathbf{Q} = (\rho, \rho u, \rho v, E)^T$, $\mathbf{F}(\mathbf{Q}) = u\mathbf{Q} + (0, p, 0, pu)^T$, $\mathbf{F}_v(\mathbf{Q}) = \mu(0, 2u_x + \lambda(u_x + v_y), v_x + u_y, u(2u_x + \lambda(u_x + v_y)) + v(v_x + u_y) + \frac{C_p}{Pr}T_x)^T$, $\mathbf{G}(\mathbf{Q}) = v\mathbf{Q} + (0, 0, p, pv)^T$, and $\mathbf{G}_v(\mathbf{Q}) = \mu(0, v_x + u_y, 2v_x + \lambda(u_x + v_y), u(v_x + u_y) +$

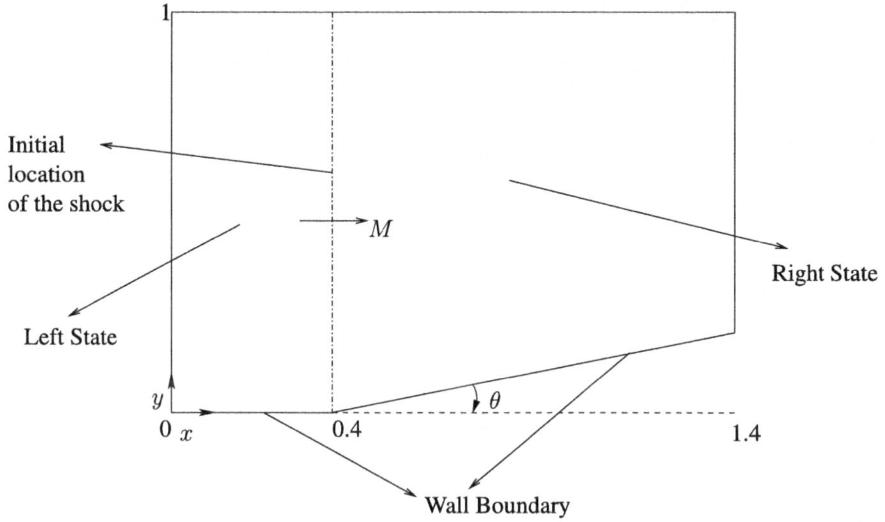

Fig. 3.1 Computational domain for a moving shock of Mach number M reflecting over a wedge of angle θ

$v(2v_y + \lambda(u_x + v_y)) + \frac{C_p}{Pr} T_y)^T$ where $p = (\gamma - 1)(E - \frac{1}{2}\rho(u^2 + v^2))$, μ is the dynamic viscosity, Pr is the Prandtl number and γ is the ratio of specific heats. By Stokes hypothesis, $\lambda = -2/3$. Here, ρ is the density, (u, v) is the velocity, E is the total energy, p is the pressure and T is the temperature. We approximate the domain Ω by K non overlapping elements given by Ω_k.

We look at solving (3.1) using the discontinuous Galerkin method. We approximate the local solution in an element Ω_k, where k is the element number, as a polynomial of order N which is given by:

$$Q_h^k(r, s) = \sum_{i=0}^{N_p - 1} Q_i^k \psi_i(r, s) \tag{3.2}$$

where $N_p = (N + 1)(N + 1)$ and r and s are the local coordinates. Here, the subscript i represents the particular degree of freedom, h represents the grid size, and the superscript k is the element number. The polynomial basis used $(\psi_i(r, s))$ is the tensor product orthonormalized Legendre polynomials of degree N. The number of degrees of freedom are given by $N_p = (N + 1)(N + 1)$. Now, using $\psi_j(r, s)$ as the test function, the weak form of the Eq. (3.1) is obtained as

$$\sum_{i=0}^{N_p - 1} \frac{\partial Q_i^k}{\partial t} \int_{\Omega_k} \psi_i \psi_j d\Omega + \int_{\partial\Omega_k} \hat{F} \psi_j ds - \int_{\Omega_k} \mathbf{F} \cdot \nabla \psi_j d\Omega = 0 \quad j = 0, \ldots, N_p - 1 \tag{3.3}$$

Fig. 3.2 Two overlapping grids (Grid 1 in black and Grid 2 in Red)

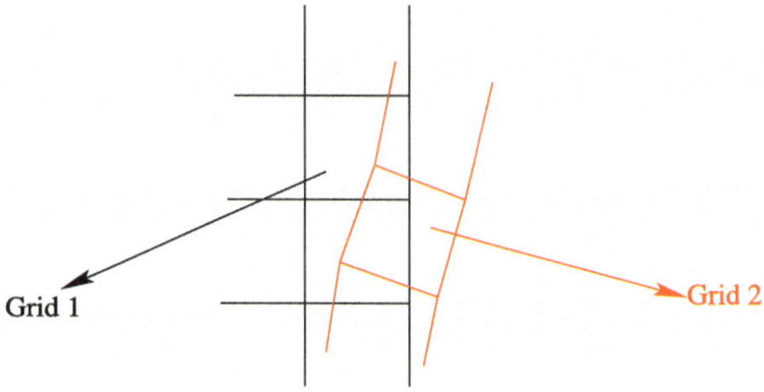

where $\partial\Omega_k$ is the boundary of Ω_k, $\mathbf{F} = (\mathbf{F}(\mathbf{Q}) - \mathbf{F}_v(\mathbf{Q}, \nabla\mathbf{Q}), \mathbf{G}(\mathbf{Q}) - \mathbf{G}_v(\mathbf{Q}, \nabla\mathbf{Q}))$ and $\hat{F} = \bar{F}^* \cdot \hat{n}$ where $\bar{F}^*$ is the monotone numerical flux at the interface which is calculated using an exact or approximate Riemann solver and $\hat{n}$ is the unit outward normal. Here, the viscous fluxes are evaluated using the BR2 method [7]. If we are solving the Euler equations, the viscous fluxes are set to zero. This is termed to be $\mathbf{P}^N$ based discontinuous Galerkin method.

Equation (3.3) is integrated using an appropriate Gauss Legendre quadrature and is discretized in time by using the fifth order Runge-Kutta time discretization given in [8] unless otherwise specified. To control spurious oscillations which occur near discontinuities, a limiter is used with a troubled cell indicator. We have used the KXRCF troubled cell indicator [10] and the compact subcell WENO (CSWENO) limiter proposed in [9] for all our calculations.

3.3 Overset Grids

Overset grids consist of multiple grids which overlap each other as shown in Fig. 3.2. When using DGM on overset grids, there are two possible approaches to handle data communication between the grids. One is a face based communication approach developed in [11], where solutions at an overset interface are obtained from the donor element, and then the boundary condition is applied weakly by imposing a numerical flux at the flux interpolation points. The other is an element based communication approach developed in [12], where the internal degrees of freedom of cells near the overset interface are obtained from the donor element. We use the new element based communication approach developed in [13] based on WENO reconstruction for the data communication between the grids.

For our problem, we use a finer overset grid near the triple point so that the flow near the triple point is captured accurately. We use a small rectangular overset grid whose location is based on the Mach number of the moving shock.

Table 3.1 Initial conditions for the three different configurations for a moving weak shock reflecting over a wedge

Configuration	M	θ	γ	$(\rho, u, v, p)_{\text{right}}$	$(\rho, u, v, p)_{\text{left}}$
1	1.47	12.5°	5/3	(5/3, 0, 0, 1)	(2.79137, 0.592296, 0, 2.45113)
2	1.075	15°	7/5	(7/5, 0, 0, 1)	(1.57697, 0.12064, 0, 1.18156)
3	1.04	11.46°	5/3	(5/3, 0, 0, 1)	(1.76663, 0.05885, 0, 1.102)

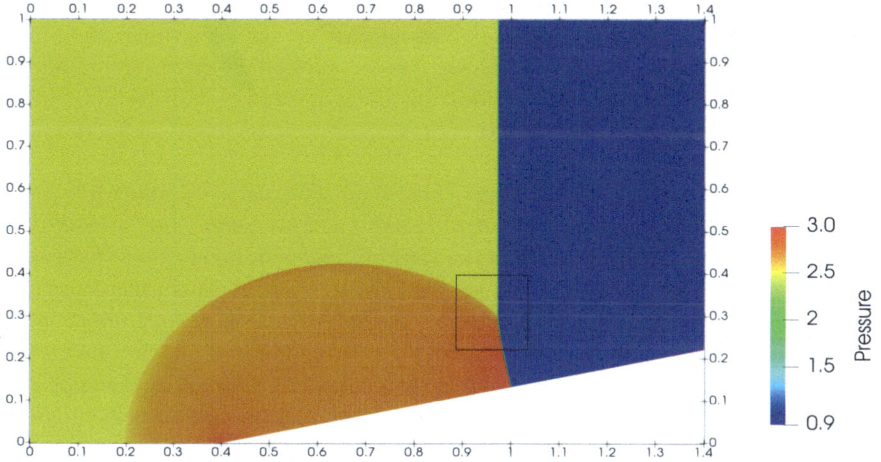

Fig. 3.3 Pressure solution obtained for Configuration - 1 ($M = 1.47$, wedge angle 12.5° and $\gamma = 5/3$) at $t = 0.32$ using $\mathbf{P}^4$ based DGM. Figure also contains a rectangular box where we use an overset grid

3.4 Results

We solve the Euler equations as given in Eq. (3.1) with the viscous fluxes set to zero for a weak shock reflecting over a wedge in the computational domain shown in Fig. 3.1 using $\mathbf{P}^4$ based DGM. We use three different configurations as the initial conditions. The initial conditions for each of the three configurations are given in Table 3.1. Configuration - 1 consists of a relatively strong weak shock while Configurations - 2 and 3 are extremely weak moving shocks. For all these three configurations, a three-shock solution does not exist. Hence all our computations are done in the von Neumann paradox regime.

For each configuration, we obtain three solutions at a given time. We first show the pressure solution obtained using a grid size of $h = 1/800$ for Configuration - 1 at $t = 0.32$ in Fig. 3.3. Figure 3.3 also contains a rectangular box containing the triple point where we propose to apply the overset grid. Now, we solve the same problem by using a fine overset grid with grid size $h = 1/1600$ to calculate the solution at the same time ($t = 0.32$). We again solve the same problem by applying viscous fluxes as given in Eq. (3.1) only in the overset grid to calculate the solution at the same time ($t = 0.32$). After obtaining these three solutions, we draw the sonic line near the triple point by locating the points where the self-similar Mach number($\bar{M}$) is 1.0 following [6]. The self-similar Mach number is defined by

$$\bar{M}^2 = \frac{(u - (x/t))^2 + (v - (y/t))^2}{c^2} \tag{3.4}$$

where c is the local speed of sound. By looking at the sonic line near the triple point we can locate the supersonic patches that occur near the reflected shock, the Mach stem and the triple point. We follow the same process for all the three configurations and show all the three solutions near the triple point at times $t = 0.32$, $t = 0.3$, and $t = 0.2$ obtained using $\mathbf{P}^4$ based DGM in Figs. 3.4, 3.5, and 3.6 for Configurations - 1, 2, and 3 respectively.

From Fig. 3.4a, we can observe a very small supersonic patch near the triple point for this configuration on grid of size $h = 1/800$. Once we keep a fine overset grid near the triple point, we observe that the number of supersonic patches have increased as shown in Fig. 3.4b. Adding the viscous fluxes on the overset grid further increases the number of supersonic patches as shown in Fig. 3.4c. Similar structures can be observed for Configurations 2 and 3. For both Configurations 2 and 3, refining the grid near the triple point (keeping an overset grid) increases the number of supersonic patches and adding viscous fluxes near the triple point further increases the number of supersonic patches. This lends credence to the observation reported in [6] that the Guderley reflection (single supersonic patch structure) is not a distinct reflection geometry and it is only an under resolved Guderley Mach reflection (multiple supersonic patch structure).

3.5 Conclusion

We have investigated the weak shock reflection problem in inviscid and viscous flows in the von Neumann paradox regime. For the three configurations we have considered, we have observed that refining the grid near the triple point increases the number of supersonic patches that occur and Guderley Mach Reflection (GMR) occurs in each case. We have also observed that the number of supersonic patches further increase if we add viscous fluxes near the triple point. These numerical solutions lend credence to the observation reported in [6] that the Guderley reflection (single supersonic patch structure) is not a distinct reflection geometry and it is only an under resolved Guderley Mach reflection (multiple supersonic patch structure).

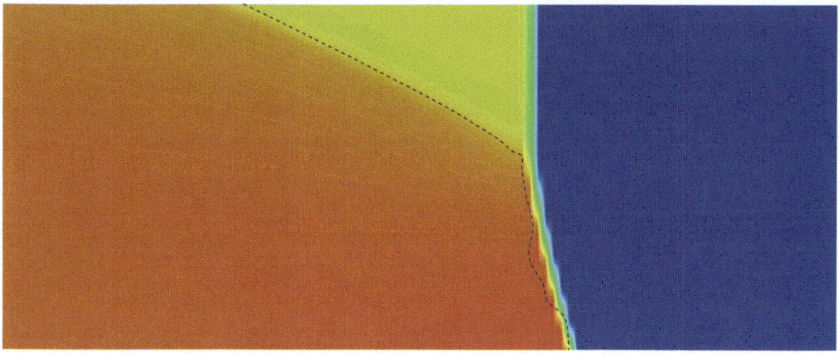

(a) Solution with $h = 1/800$

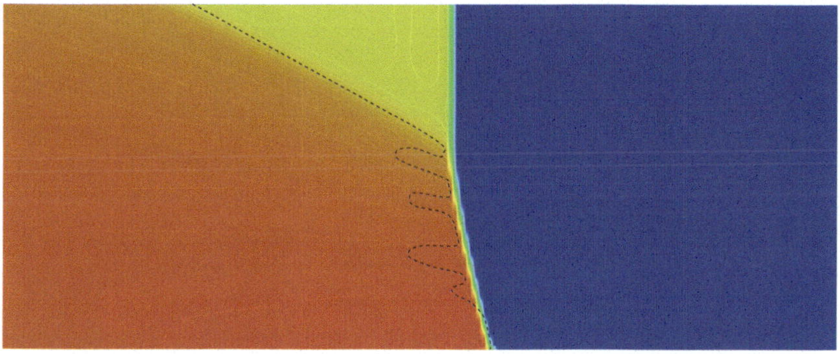

(b) Solution with $h = 1/800$ and an overset grid near triple point

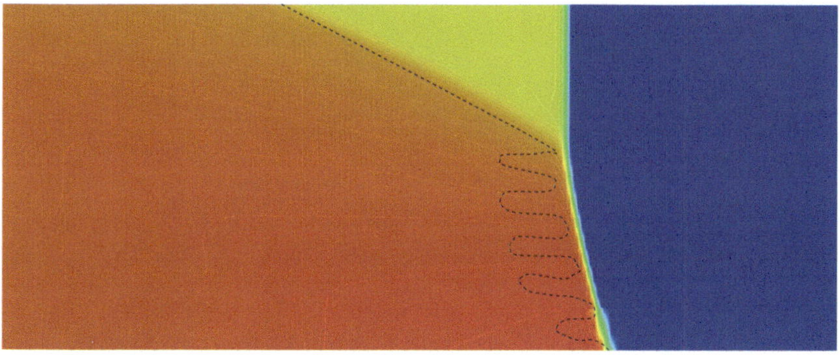

(c) Solution with $h = 1/800$ with viscous fluxes applied on overset grid

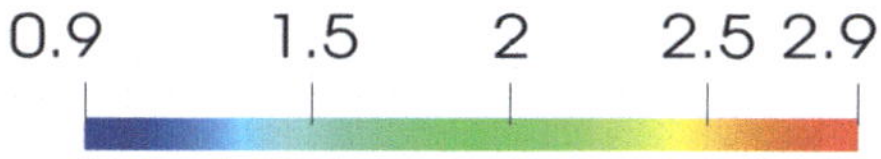

(d) Pressure Range

Fig. 3.4 Pressure contours along with the sonic line (black dashed line) for Configuration - 1 ($M = 1.47$, wedge angle 12.5° and $\gamma = 5/3$) showing the sonic line for three cases - 1)$h = 1/800$ grid solution, 2) $h = 1/800$ grid and an overset grid of $h = 1/1600$ near triple point, and 3)$h = 1/800$ grid and an overset grid of $h = 1/1600$ near triple point with viscous fluxes applied only on the overset grid obtained using $\mathbf{P}^4$ based DGM

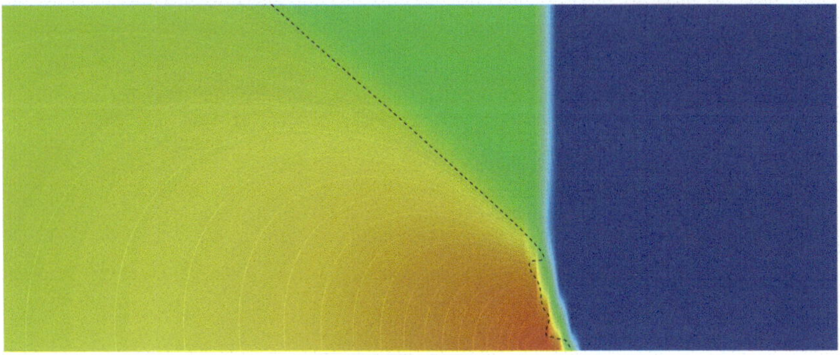

(a) Solution with $h = 1/800$

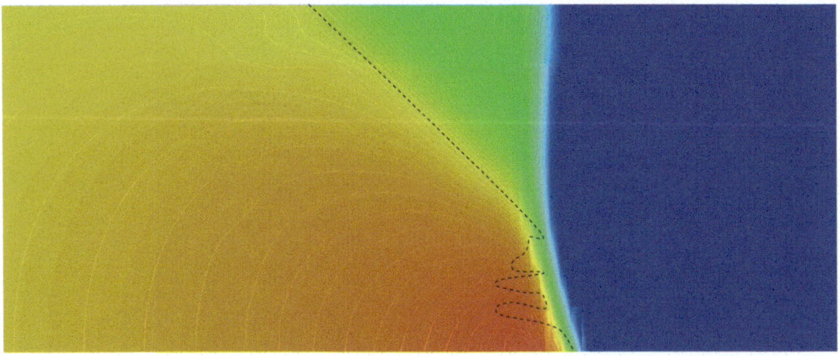

(b) Solution with $h = 1/800$ and an overset grid near triple point

(c) Solution with $h = 1/800$ with viscous fluxes applied on overset grid

(d) Pressure Range

Fig. 3.5 Pressure contours along with the sonic line (black dashed line) for Configuration - 2 ($M = 1.075$, wedge angle 15° and $\gamma = 7/5$) showing the sonic line for three cases - 1)$h = 1/800$ grid solution, 2) $h = 1/800$ grid and an overset grid of $h = 1/1600$ near triple point, and 3)$h = 1/800$ grid and an overset grid of $h = 1/1600$ near triple point with viscous fluxes applied only on the overset grid obtained using $\mathbf{P}^4$ based DGM

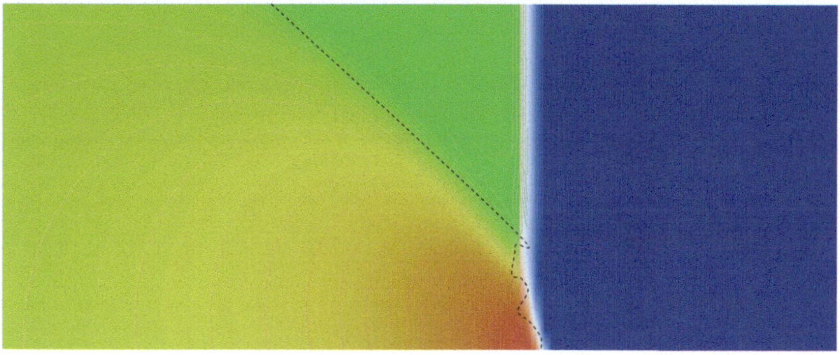

(a) Solution with $h = 1/800$

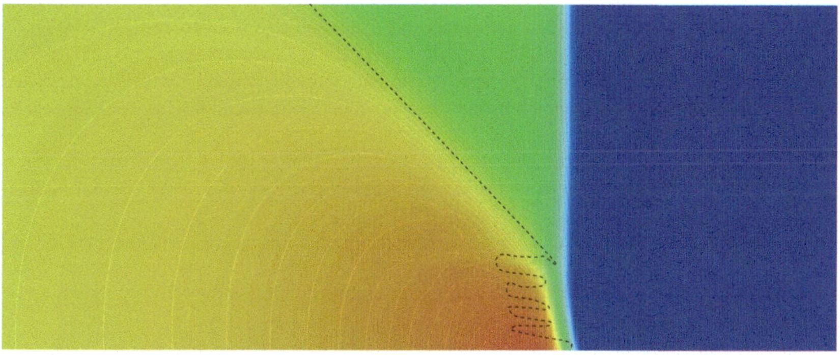

(b) Solution with $h = 1/800$ and an overset grid near triple point

(c) Solution with $h = 1/800$ with viscous fluxes applied on overset grid

(d) Pressure Range

Fig. 3.6 Pressure contours along with the sonic line (black dashed line) for Configuration - 3 ($M = 1.04$, wedge angle 11.46° and $\gamma = 7/5$) showing the sonic line for three cases - 1)$h = 1/800$ grid solution, 2) $h = 1/800$ grid and an overset grid of $h = 1/1600$ near triple point, and 3)$h = 1/800$ grid and an overset grid of $h = 1/1600$ near triple point with viscous fluxes applied only on the overset grid obtained using $\mathbf{P}^4$ based DGM

References

1. W. Bleakney, A. Taub, Interaction of shock waves. Rev. Modern Phys. **21**, 584–605 (1949)
2. K.G. Guderley,The theory of transonic flow. Pergamon press (1962)
3. E. Vasil'ev, A. Kraiko, Numerical simulation of weak shock diffraction over a wedge under the von Neumann paradox conditions. Comput. Math. Math. Phys. **39**, 1335–1345 (1999)
4. A.M. Tesdall, J.K. Hunter, Self-similar solutions for weak shock reflection. SIAM J. Appl. Math. **63**, 42–61 (2002)
5. B. Skews, G. Li, R. Paton, Experiments on Guderley Mach reflection. Shock Waves **19**, 95–102 (2009)
6. A.M. Tesdall, R. Sanders, N. Popivanov, Further results on Guderley mach reflection and the triple point paradox. J. Sci. Comput. **64**, 721–744 (2015)
7. F. Bassi, A. Crivellini, S. Rebay, M. Savini, Discontinuous Galerkin solution of the Reynolds-averaged Navier-Stokes and k-ω turbulence model equations. Comput. Fluids **34**, 507–540 (2005)
8. J.C. Butcher, Numerical methods for ordinary differential equations. 3rd edn., John Wiley and Sons (2016)
9. S.R. Siva Prasad Kochi, M. Ramakrishna, A compact subcell WENO limiting strategy using immediate neighbours for Runge-Kutta discontinuous Galerkin methods. Int. J. Comput. Math. **98**(3), 608–626 (2021)
10. L. Krivodonova, J. Xin, J.-F. Remacle, N. Chevaugeon, J. Flaherty, Shock detection and limiting with discontinuous Galerkin methods for hyperbolic conservation laws. Appl. Numer. Math. **48**, 323–338 (2004)
11. M.C. Galbraith, J.A. Benek, P.D. Orkwis, M.G. Turner, A discontinuous Galerkin Chimera scheme. Comput. Fluids **98**, 27–53 (2014)
12. C.R. Nastase, D.J. Mavriplis, J. Sitaraman, An overset unstructured mesh discontinuous Galerkin approach for aerodynamic problems. AIAA 2011-195 (2011)
13. S.R. Siva Prasad Kochi, M. Ramakrishna, A. discontinuous Galerkin overset scheme using WENO reconstruction and subcells for two-dimensional problems. J. Sci. Comput. **93**, 35 (2022)

Chapter 4
Numerical Investigations of the Effect of Unsteady Generator Pressure on Supersonic Ejector Performance

C. V. Sumesh Babu, P. J. Manoj, A. J. Leo, V. Nikhil Jacob, and V. Lijo

Abstract Ejector Refrigeration systems (ERS) is experiencing a renewed interest compared to the conventional refrigeration system, owing to the significant reduction in electricity demand and rise in awareness of ozone depletion. The preferred mode of operation which gives maximum entrainment ratio (ER) is choking of both primary and secondary streams at different locations, known as double-choking (critical mode). High values of ER show greater carrying capacity of the motive fluid and hence improved pumping performance. The substantial fluctuating behavior of driving energy for the ejector may force ejector operation away from the narrow range around design conditions and can severely affect the performance of ERS. The present work considers unsteady numerical modeling of a supersonic ejector with generator pressure variations while maintaining constant suction and discharge pressures. The investigation reveals that as the primary stagnation pressure builds, the ejector operation mode transits from single choked to double choked, and ER reaches its optimum at the point of transition. The evolution Fabri-choke, visualization of Fabri-choke, associated complex shock structures, and their relation to ER are discussed in detail.

4.1 Introduction

Owing to the demand for human thermal comfort, ejector refrigeration systems (ERS) experienced a renewed interest since they could use energy sources like solar energy, geothermal energy, or waste heat energy, thus, minimizing the electricity requirement. In ERS, a supersonic ejector replaces the conventional compressor for the compression process. Moreover, ERS avoids harmful emissions of refrigerants which cause

C. V. Sumesh Babu (✉) · P. J. Manoj · A. J. Leo · V. Nikhil Jacob · V. Lijo
Department of Mechanical Engineering, Government Engineering College Thrissur, Thrissur, Kerala 680009, India
e-mail: sumesh.bcv@gmail.com

APJ Abdul Kalam Technological University, Thiruvananthapuram, Kerala, India

© The Author(s), under exclusive license to Springer Nature Singapore Pte Ltd. 2024
G. Rajesh et al. (eds.), *Advances in Shock Interactions*, Springer Proceedings in Physics 408, https://doi.org/10.1007/978-981-97-6099-2_4

the greenhouse effect that leads to global warming. A supersonic ejector is a non-moving mechanical device that consists of a supersonic primary nozzle, secondary nozzle, mixing duct, and subsonic diffuser. In ERS high energy motive force called primary fluid is produced in the generator (boiler). The primary fluid is then allowed to pass through the convergent-divergent primary nozzle to attain a supersonic speed at the nozzle exit position (NXP). The low-pressure region developed due to expansion of this primary jet downstream of NXP creates a favorable condition for the entrainment of secondary fluid from the evaporator. The required compression (back-pressure) is achieved by the oblique shock interaction, pseudo shock, and subsonic compression in the diffuser. In the present study, the authors are particularly interested in solar powered ERS.

The performance of supersonic ejectors is expressed in terms of entrainment ratio (ER), which is the ratio of secondary mass flow rate to the primary mass flow rate. The schematic of the variation of ER with primary pressure (called generator pressure in the present paper) is shown in Fig. 4.1 for a given value of suction pressure and backpressure. At lower values of generator pressure (P_{0p}), straight shock produced inside the primary nozzle makes the entire flow in the ejector subsonic, and the information on backpressure is transmitted to the entire flow [2]. Moreover, the kinetic energy exchange between the subsonic primary fluid (primary fluid becomes entirely subsonic because of the strong shock right after the downstream of NXP) and the secondary fluid is sufficiently low [2]. Hence, when the generator pressure is less than a threshold value called break-down pressure, the backflow (flow reverses towards the secondary nozzle) takes place instead of entrainment. For pressure higher than break-down pressure, ER is very sensitive to the P_{0p} and increases almost linearly [2]. In this stage, as the generator pressure increases, the jet becomes more under-expanded downstream of NXP. It thus creates a favorable vacuum pressure for entrainment of the secondary fluid, and the ejector operation at this stage is termed as sub-critical mode [1]. In the sub-critical mode, entrainment of secondary fluid begins with low values of subsonic speed, which increases with P_{0p}, and eventually reaches the sonic velocity, and the flow chokes (ER also reaches the maximum value). The secondary fluid choking was first introduced by Fabri and Siestrunck [3] in their so-called aerodynamic theories, and hence it is referred to as Fabri-choke in this paper. The P_{0p} at which the ER maximizes is called critical primary pressure [4], and the ejector operation beyond P_{0p} is termed as critical mode.

The double choked mode of ejector operation is not common in all sets of geometries or operating conditions. The given ejector geometry allows only a narrow range of optimal operating conditions under which the ER becomes optimum [5]. Figure 4.1 shows that if we increase the P_{0p} beyond critical generator pressure, ER decreases. Thus, finding the value of critical P_{0p} is of utmost importance otherwise, that will lead to the wastage of primary fluid energy [4]. Although several researchers put forward a few reasons behind the cause of ER reduction [2, 5], as the P_{0p} increases beyond the critical value, those were not specific and demanded further investigation.

In all the practical applications of ERS, the operating conditions are highly fluctuating in nature depending on the availability of waste heat or solar energy, cooling load demand, and evaporator conditions. Conducting the transient analysis is the

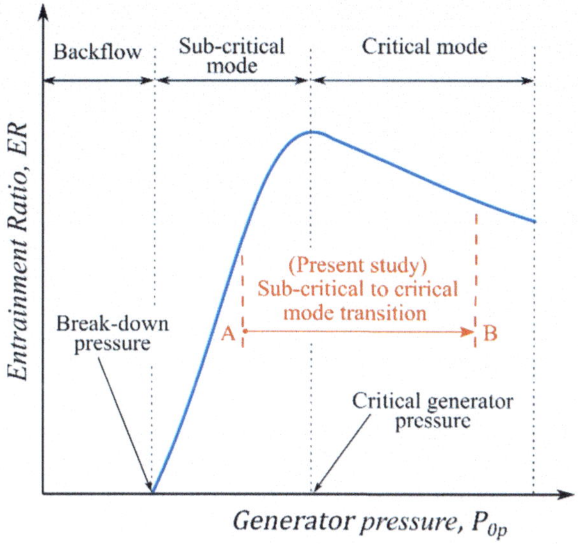

Fig. 4.1 Schematic of variation of ER with generator pressure [1]

best remedy to simulate the actual behavior of supersonic ejectors. Few researchers conducted the transient analysis [6] on supersonic ejectors, but their investigations were on vacuum ejectors. Bartosiewicz et al. [2] introduced the concept of sonic line plotting in the numerical investigation of a supersonic ejector. Hemidi et al. [7] employed the sonic line method to compare the turbulence models and suggested that the location of critical choke point can be ascertained with sonic line potting. Lamberts et al. [8] argued the visualization of critical choke (Fabri-choke) with the help of sonic line plotting for the first time, but the reported data is of steady state only. Recently Sumesh Babu and Lijo [9] conducted the transient analysis to visualize Fabri-choke in a supersonic ejector, but the work was carried out at a constant primary pressure only. Hence, in the present paper, we have conducted a transient numerical analysis by varying generator pressure (line AB in Fig. 4.1) to investigate the evolution FC in relation with movement of shock structure in the entrance of mixing duct, and to discover the variation of ER while the ejector operation switches from sub-critical mode to critical mode.

4.2 Computational Method

The schematic of the ejector (Fig. 4.2) for the present numerical study is adapted from the experimental investigation of mixing length study of supersonic ejectors [10]. The experiment ejector is of a rectangular type having a width of 80 mm; hence, a 2D computation is adopted. A convergent-divergent primary nozzle is placed inside a secondary nozzle. The design Mach number of the primary nozzle is 2.0 with a throat of 3.6 mm. The primary NXP is placed 15 mm upstream of the mixing duct inlet. The height of the constant area mixing duct is 20 mm, and its length is 200 mm. The secondary inlet has a height of 60 mm and converges to the mixing duct height of 20 mm. A subsonic diffuser of 200 mm in length and exit height of 60 mm is provided downstream of the mixing duct. In the present numerical study, static pressures at 41 locations (5 mm spacing) in the mixing duct wall are monitored to capture the exact location of occurrence of FC and marked as MP1, MP2, … MP41. The axial distances are measured from the primary nozzle inlet (x = 0). A constant value of secondary stagnation pressure (P_{0s}) of 0.86 bar and the condenser pressure of 2.5 bar is provided. Stagnation temperature is taken as 300 K for both primary and secondary streams. To begin with, steady-state analysis was conducted for the ejector at a relatively low generator pressure of 9 bar, which is in sub-critical mode (single choking). This steady-state solution is then taken as the initial condition for the unsteady simulations with an assumed linear increment in generator pressure from 9 to 11 bar (Fig. 4.4). Such an increment will change the operational mode of the ejector from sub-critical mode to critical mode.

Bartosiewicz et al. [11] compared the experimental model with six turbulence models (standard *k*-*ε*, realizable *k*-*ε*, standard *k*-*ω*, RSM, RNG, and *SST k*-*ω*) and reported that, *SST k*-*ω* model is best suited to obtain the correct ER, predict the shock position, shock boundary layer interaction, and static pressure recovery and hence *SST k*-*ω* is adopted for the present investigation. Two-dimensional, transient numerical simulations were performed using the commercial *CFD* software *ANSYS* Fluent16. Dry air with ideal gas assumption is taken as the working substance used in the present study with viscosity calculated by Sutherland law. All the walls were considered to be adiabatic with no-slip. The mesh consists of 0.3 million purely structured quad elements, and the time-step selected is 5×10^{-8} s after grid and time-step independence studies. The minimum grid space normal to the wall is 10^{-6} m ($y^+ \sim 1.0$).

Figure 4.3 compares the experimentally observed variations in wall static pressure with the code for a fixed backpressure (p_b) of 1.34 bar. Numerical modeling has

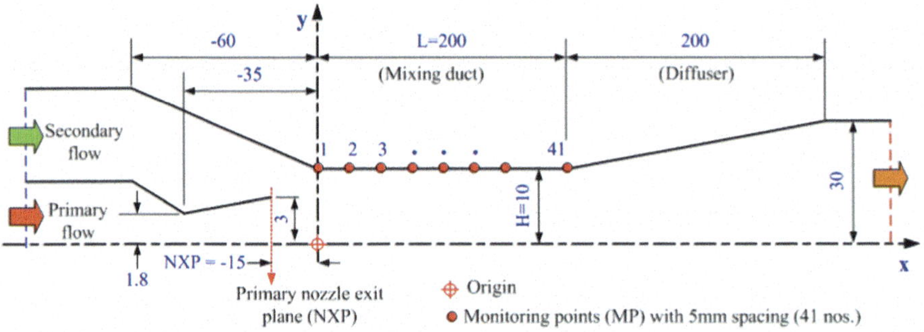

Fig. 4.2 Schematic of the 2D model ejector along with pressure monitoring points used in the simulation (all dimensions in mm)

been done with the exact geometry and flow conditions as in the actual 2D supersonic ejector [10]. The primary pressure (P_{0p}) observed during the experiment was 7.69 bar. During the operation of the ejector under the above-said conditions, an entrainment ratio (ER) of 0.44 was achieved. The agreement between the predicted wall pressure distribution and the experimental values is good and demonstrates the proper behavior.

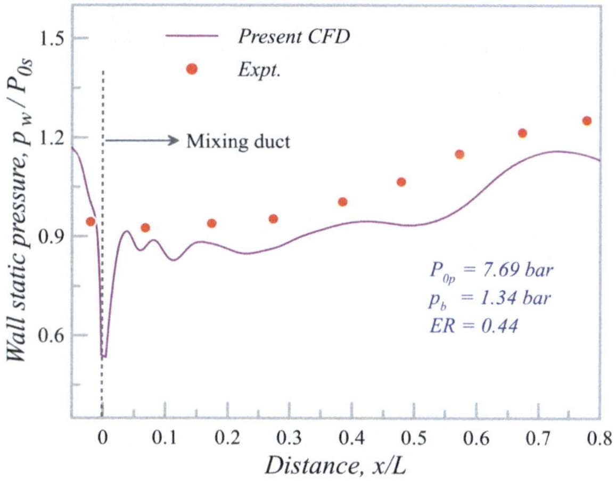

Fig. 4.3 Numerical validation with the experimental results

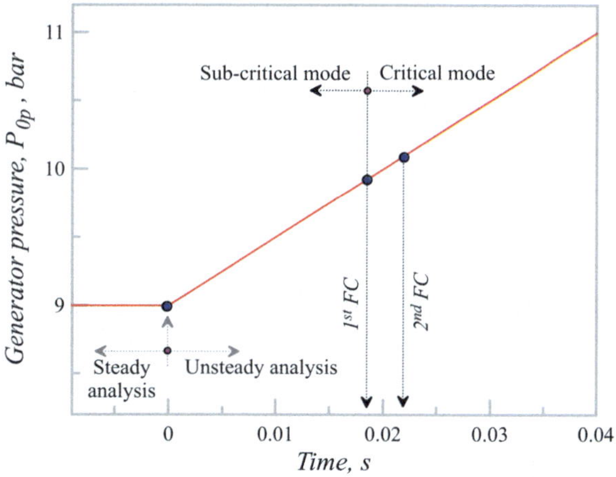

Fig. 4.4 Generator pressure variation with time selected for unsteady simulation

4.3 Results and Discussions

Initially, steady state analysis was conducted by keeping a constant value of P_{0p} at 9 bar, ensuring that the ejector is at the sub-critical mode of operation. Transient analysis has been carried out from the fields generated from the above steady state solution by increasing the P_{0p} from 9 to 11 bar (linear variation was assumed, Fig. 4.4) in 40 ms, keeping constant values for all other operating conditions. The operation mode of the ejector has been switched from sub-critical to critical during the analysis.

Figure 4.5 shows the entrainment ratio history which indicates that ER initially increases with the increase of P_{0p} in the sub-critical mode of ejector operation. In the sub-critical mode, on increasing the P_{0p}, ER reaches its maximum value during the first occurrence of Fabri-choke at t = 18.4 ms. The ER decreases with a further increase of the P_{0p} in the critical mode.

Sonic line (M = 1) distribution which demarcates the subsonic and supersonic regions, is an excellent tool for the numerical visualization of Fabri-choke formation [2, 7]. Figure 4.6 shows the sonic line distribution of the various instants of ejector operation in the subcritical mode. In the subcritical mode, as the P_{0p} increases, the primary jet undergoes more under expansion (indicated by the growth of sonic line) near the entrance of mixing duct.

This under-expansion growth creates a low pressure for the entrainment of secondary fluid along with the increase of secondary flow speed. The speed of secondary flow increases with the increase in P_{0p} and reaches the sonic condition and chokes (1st Fabri-choke) at t = 18.4 ms. The Fabri-choke formation is indicated by the sudden enlargement of sonic line in the lateral direction. The ER also reaches its maximum value along with the first occurrence of Fabri-choke, and the ejector operation turns to the critical mode.

Figure 4.7 shows the sonic line distribution at the entrance of mixing duct at various instants in the critical mode of operation of the ejector. In the critical mode,

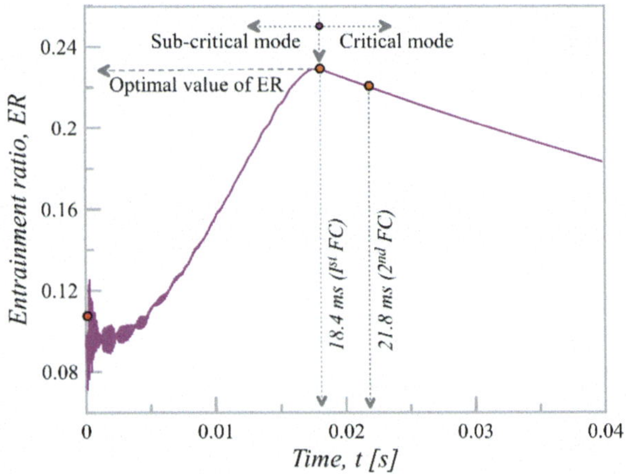

Fig. 4.5 Entrainment ratio history

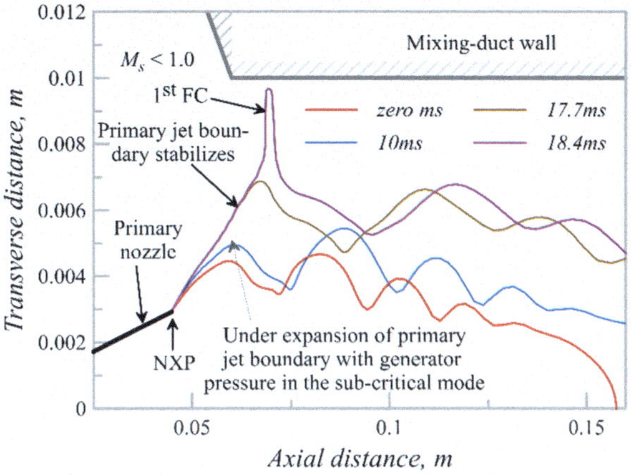

Fig. 4.6 Sonic line development at mixing duct entrance on the sub-critical mode of ejector

the growth of the primary jet stabilizes and remains unaltered with the increase of the P_{0p}. This ceasing of under-expansion increases static pressure at the mixing duct entrance so that suction pressure for the entrainment of secondary flow decreases, and hence, the ER decreases. Also, the supersonic region downstream of the first FC location enlarges with the increase of generator pressure. Moreover, another choking point (second occurrence of FC) is also formed downstream at t = 21.8 ms.

Figure 4.8 shows the pressure history of the monitoring points MP1 (60 mm, beginning of the mixing duct) and MP3 (70 mm, location of 1st FC at t = 18.4 ms). Initially, for the monitoring point MP1, on increasing P_{0p}, the wall static pressure decreases due to the growth of under-expansion and reaches a minimum value, providing a favorable pressure gradient for the secondary fluid to entrain; hence, the ER increases accordingly. Further increasing of P_{0p} (primary jet under expansion stabilized), the wall static pressure is observed to increase, and the ER decreases. As the wall static pressure is not breaching the critical choking pressure (1D theory) for MP1, the

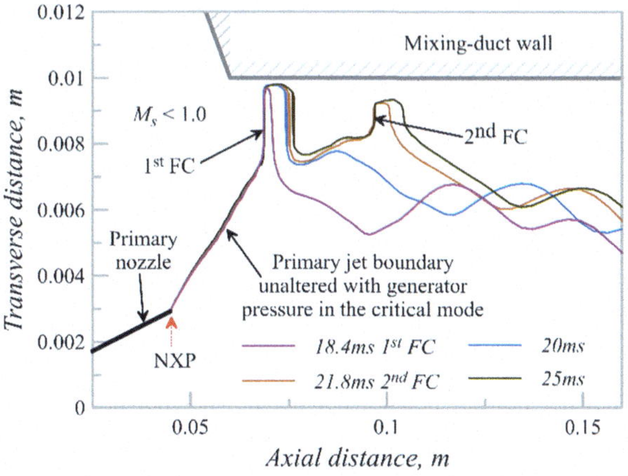

Fig. 4.7 Sonic line development at mixing duct entrance on the critical mode of ejector

secondary flow remains subsonic throughout. Shortly after that, at MP3, the static pressure falls below the critical choking limit at time t = 18.4 ms, and the subsonic secondary flow turns supersonic (at MP3), and chokes; the ER attains its maximum value.

For the quantitative visualization of secondary flow choking, Mach number distribution at the lateral section of the first FC location is taken in Fig. 4.9. The figure shows the primary-jet core, the shear layer (where momentum exchange takes place), and the secondary flow region near the mixing duct wall. At time t = 17.7 ms, the secondary flow is not gained the sonic speed, but at t = 18.4 ms, the secondary flow becomes sonic and chokes. Moreover, at t = 40 ms, secondary flow changed to subsonic, indicating that the FC location is shifted downstream (Fig. 4.10) as the P_{0p} increases.

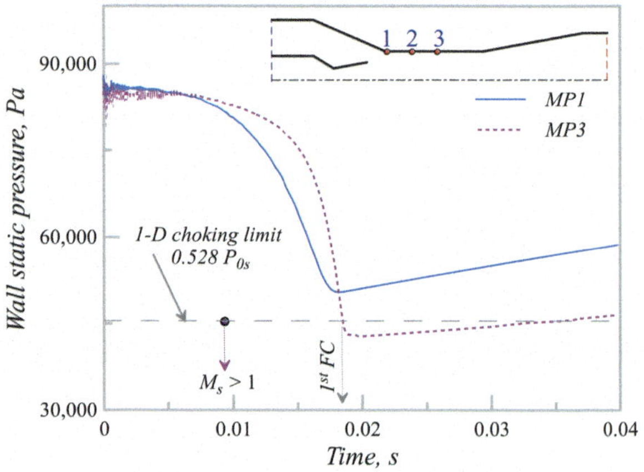

Fig. 4.8 Static pressure histories of the monitoring points

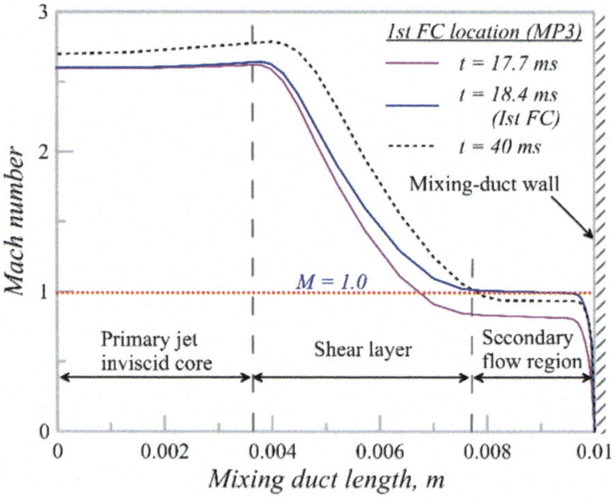

Fig. 4.9 The lateral distribution of Mach number inside the mixing duct at section MP3

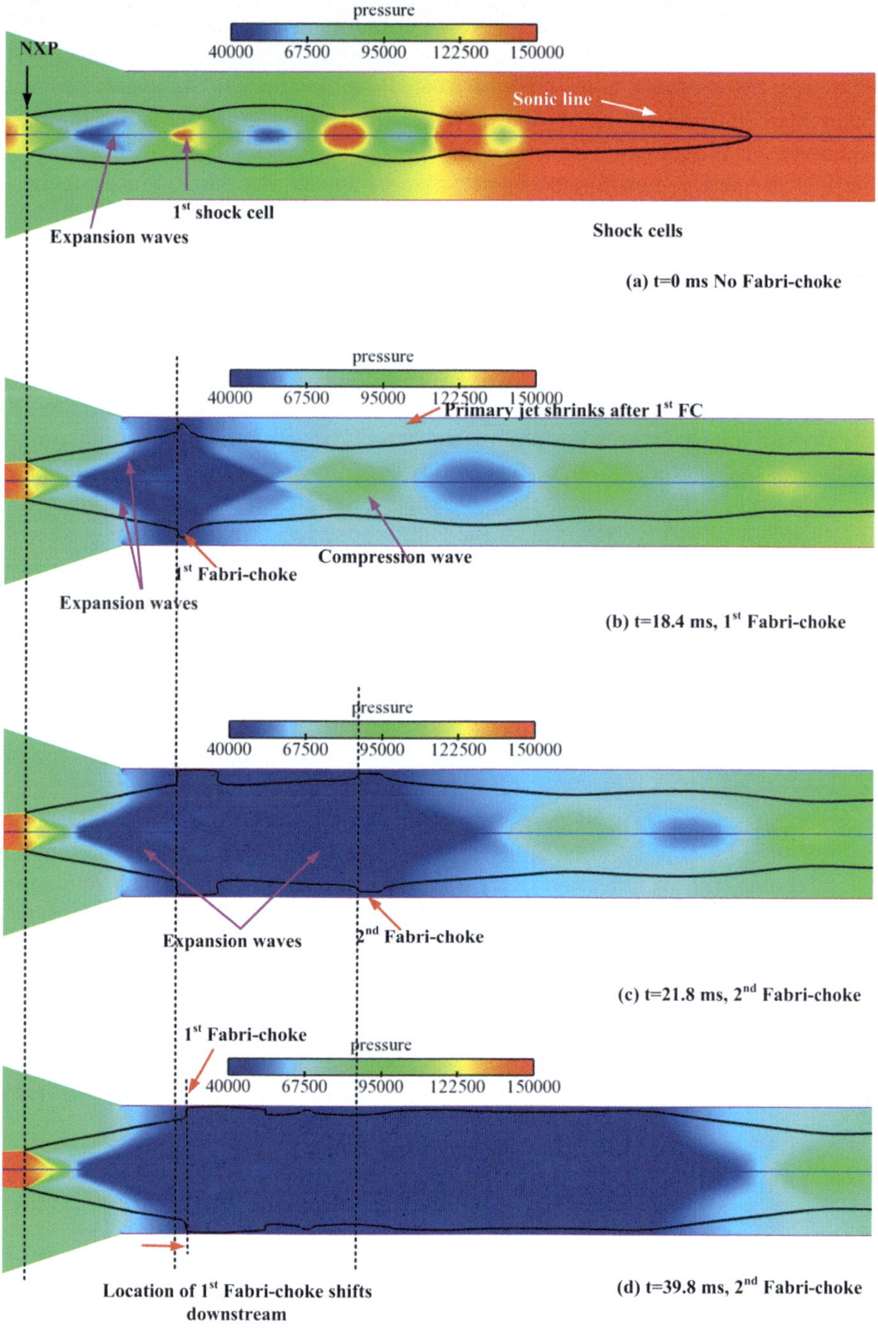

Fig. 4.10 Static pressure contours showing the formation of FC with wave reflections

We have investigated the occurrence of FC and its association with wave reflections (both expansion and shock waves) which is made clear with instantaneous static-pressure contours with overlaid sonic lines (Fig. 4.10). At t = 0 ms, the under-expanded primary jet shows alternate expansion and compression regions. Expansion waves from the nozzle exit impinge on the jet boundary creating a favorable low-pressure region for the Fabri-choke to appear. As the P_{0p} increases at t = 18.4 ms, expansion waves from the nozzle exit strengthen, which get reflected from jet boundary as compression waves (indicated by shrinking of the jet boundary after first FC). Interestingly, the first FC occurs precisely where the expansion wave reflects from the jet boundary. With further increase of P_{0p} at t = 21.8 ms and 40 ms (Fig. 4.10), downstream of the first FC, expansion waves are formed as the primary jet enlarges. These expansion waves impinging on the jet boundary create a low-pressure region aiding the formation of multiple choke points downstream. A shockless region can be seen upstream of the first FC (Fig. 4.10), which further moves downwards considerably with the formation of multiple Fabri-chokes. Although the shockless region extended up to the half length of mixing duct, that does not favor the enhancement of ER. However, at the Fabri-choke location, sudden enlargement of the sonic line decreases the annular area for the secondary flow to pass through (funneling effect theory by Munday and Bagster [12]), which minimizes the contact area available for the two fluids to interact [2], prevent the further enhancement of the ER.

Instantaneous numerical schlieren to address the complex system of expansion and shock waves during the formation of FC is also investigated and detailed in Fig. 4.11. At t = 0 ms (no FC; Fig. 4.11), the intercepting shocks undergo a regular type of shock reflection (RR) only. With further increase of P_{0p} the primary jet becomes more and more under-expanded, resulting in strong wave reflections. During the first occurrence of FC (t = 18.4 ms, Fig. 4.11), the intercepting shock became so intense that a regular type of reflection (RR) was impossible, and hence Mach reflection (MR) took place. The occurrence of the Mach reflection and its interaction with the secondary fluid turns the secondary flow to subsonic (at the Mach stem).

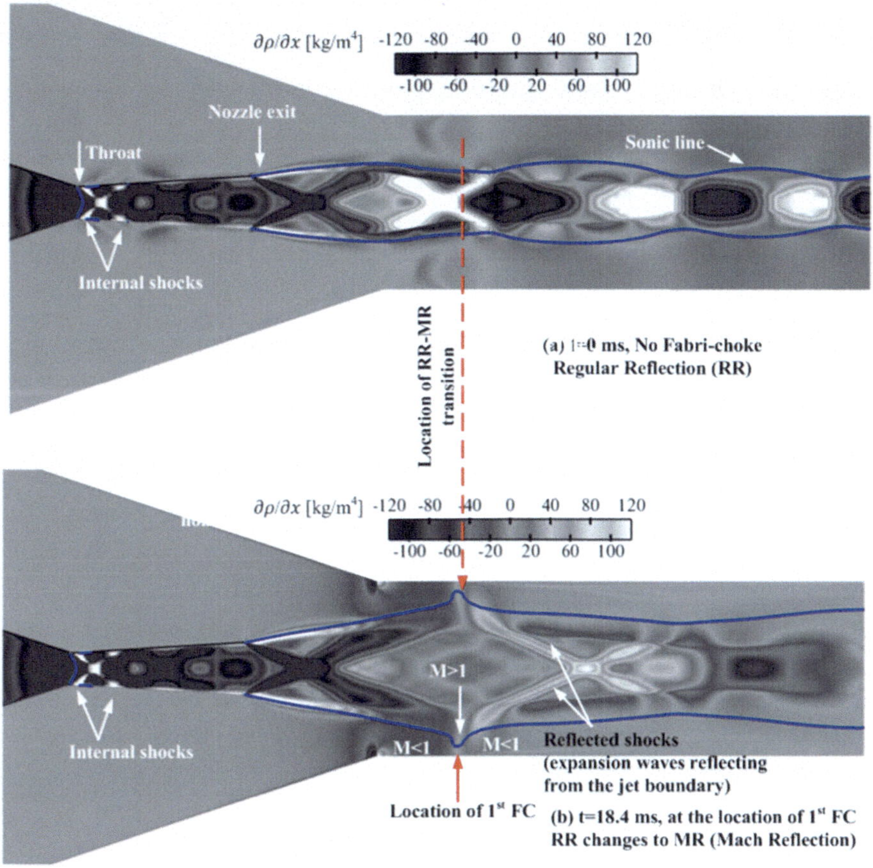

Fig. 4.11 Numerical schlieren showing the types of shock wave reflection in the region of FC

4.4 Conclusions

The present work numerically investigates the effect of unsteady generator pressure in a supersonic ejector typically used in an ejector refrigeration cycle. A linear transient variation in generator pressure is selected for the study, so the ejector operation switches from sub-critical to critical mode. The occurrence of FC can be considered as an ejector operating in critical mode. The ER initially increases with P_{0p} as the primary jet undergoes under expansion and gets optimized when the operation mode of the ejector switches to critical mode due to the stabilization of under expansion of the primary jet. The wall static pressure at the entrance of the mixing duct increases with a further increase of P_{0p}, reducing ER. Although the shockless region extends downstream, with increase of P_{0p}, the contact surface for the interaction of two fluids minimizes during the formation of FC, reducing ER. Fabri-choke is formed precisely where the expansion wave from the nozzle exit hits the primary jet boundary. Also, regular reflection to Mach Reflection transition takes place at the location of FC.

References

1. G. Besagni, R. Mereu, F. Inzoli, Ejector refrigeration: a comprehensive review. Renew. Sustain. Energy Rev. **53**, 373–407 (2016). https://doi.org/10.1016/j.rser.2015.08.059
2. Y. Bartosiewicz, Z. Aidoun, P. Desevaux, Y. Mercadier, Numerical and experimental investigations on supersonic ejectors. Int. J. Heat Fluid Flow **26**(1), 56–70 (2005). https://doi.org/10.1016/j.ijheatfluidflow.2004.07.003
3. J. Fabri, R. Siestrunck, Supersonic air ejectors. Adv. Appl. Mech. **5**, 1–34 (1958). https://doi.org/10.1016/S0065-2156(08)70016-4
4. L. Cai, H.T. Zheng, Y.J. Li, Z.M. Li, Computational fluid dynamics simulation of the supersonic steam ejector. Part 2. Optimal design of geometry and the effect of operating conditions on the ejector. Proc. Inst. Mech. Eng. Part C J. Mech. Eng. Sci. **226**(3), 715–723 (2012). https://doi.org/10.1177/0954406211415779
5. Z. Aidoun, M. Ouzzane, The effect of operating conditions on the performance of a supersonic ejector for refrigeration. Int. J. Refrig. **27**(8), 974–984 (2004). https://doi.org/10.1016/j.ijrefrig.2004.05.006
6. V. Lijo, H.D. Kim, T. Setoguchi, S. Matsuo, Numerical simulation of transient flows in a rocket propulsion nozzle. Int. J. Heat Fluid Flow **31**(3), 409–417 (2010). https://doi.org/10.1016/j.ijheatfluidflow.2009.12.005
7. A. Hemidi, F. Henry, S. Leclaire, J.M. Seynhaeve, Y. Bartosiewicz, CFD analysis of a supersonic air ejector. Part II: relation between global operation and local flow features. Appl. Therm. Eng. **29**(14–15), 2990–2998 (2009). https://doi.org/10.1016/j.applthermaleng.2009.03.019
8. O. Lamberts, P. Chatelain, Y. Bartosiewicz, Numerical and experimental evidence of the Fabri-choking in a supersonic ejector. Int. J. Heat Fluid Flow **69**, 194–209 (2018). https://doi.org/10.1016/j.ijheatfluidflow.2018.01.002
9. C. V. Sumesh Babu, V. Lijo, A computational study of the evolution of Fabri-choke in a two-dimensional supersonic ejector, in *ASME International Mechanical Engineering Congress and Exposition IMECE*, (2021), p. IMECE2021-70919
10. S.M.V. Rao, G. Jagadeesh, Observations on the non-mixed length and unsteady shock motion in a two dimensional supersonic ejector. Phys. Fluids **26**(3), 036103 (2014). https://doi.org/10.1063/1.4868879

11. Y. Bartosiewicz, Z. Aidoun, P. Desevaux, Y. Mercadier, CFD-experiments integration in the evaluation of six turbulence models for supersonic ejector modeling, in *Integrating CFD and Experiments Conference,* Glasgow, UK, no. 450, (2003), [Online]. Available: http://www.cfd4aircraft.com/int_conf/IC1/papers/Bartosiewicz.pdf
12. J.T. Monday, D.F. Bagster, A new ejector theory applied to steam jet refrigeration. Ind. Eng. Chem. Process. Des. Dev. **16**(4), 442–449 (1977). https://doi.org/10.1021/i260064a003

Chapter 5
Temperature Calibration of a Single Pulse Shock Tube at Low Pressures Using Unimolecular Decomposition of Cyclohexene

Chaitanya Bhoir, Deepak Kumar Singh, Elangannan Arunan, and Jagadeesh Gopalan

Abstract This paper presents the chemical thermometry of a single pulse shock tube with cyclohexene as an external standard to measure the actual temperature behind the reflected shock wave ($T_{5,act}$) under low-pressure conditions. The temperature behind the reflected shock wave was determined using the initial and final concentration of cyclohexene acquired from the experiments, reaction dwell time and the rate constant expression given in the literature. The shock tube is calibrated in the temperature ($T_{5,act}$) range of 970–1136 K, and this calibration is valid to the pressure range of 0.81–3.75 bar. The theoretical temperature behind the reflected shock wave ($T_{5,theo}$) was computed from the 1-D normal shock relations using the incident shock wave's end wall Mach number and the initial temperature in the driven section. Finally, the linear correlation has been established between the theoretical and actual temperature behind the reflected shock wave for chemical shock tube (CST-3) present in our lab. These experiments highlight the notable difference in theoretical and actual temperature, which increases with the increase in temperature, underscoring the need for shock tube calibration before ignition delay time measurements and other kinetic studies. The drawback of using cyclohexene's unimolecular decomposition as a chemical thermometer above a specific temperature has also been discussed.

C. Bhoir · J. Gopalan (✉)
Department of Aerospace Engineering, Indian Institute of Science, Bengaluru 560012, India
e-mail: jaggie@iisc.ac.in

D. K. Singh · E. Arunan
Department of Inorganic and Physical Chemistry, Indian Institute of Science, Bengaluru 560012, India

© The Author(s), under exclusive license to Springer Nature Singapore Pte Ltd. 2024
G. Rajesh et al. (eds.), *Advances in Shock Interactions*, Springer Proceedings in Physics 408, https://doi.org/10.1007/978-981-97-6099-2_5

5.1 Introduction

The ignition delay time (IDT) of the fuels plays a crucial role in designing an engine. Over the years, the shock tube has proven to be a reliable tool for obtaining IDT data of different fuels under a wide range of operating conditions [1]. In addition to IDT data, product species concentration obtained after ignition delay experiments in the shock tube can be utilized to validate the reduced kinetic mechanism for fuel combustion. Integration of these accurate reduced kinetic mechanisms with computational fluid dynamics (CFD) can assist in optimizing the engine's design without the need for costly experiments, therefore saving enormous cost and time. Because of its potential applications, such as hypersonic cruise missiles and more affordable entry into space, the scramjet engine has been under focus for many decades [2]. In the scramjet engine, supersonic combustion occurs near atmospheric pressures [3]; hence investigating the IDT and developing an accurate reduced kinetic mechanism of various fuels in that pressure range is essential in its design and development.

We are aiming to generate IDT data and accurate reduced kinetic mechanism for several jet fuels in low-pressure range as it is lacking presently. However, the data produced during the experiments is only meaningful if we know the actual operating conditions during the measurements. In our studies, the temperature behind the reflected shock wave is very crucial parameter because quantities like IDT and reaction rate strongly depends on it. Therefore, knowing the actual temperature behind the reflected shock wave is essential for generating reliable data.

5.1.1 Single Pulse Shock Tube

A shock tube is basically a long cylindrical tube with closed ends divided into two sections with the help of a diaphragm [4]. One section of the tube is filled with high-pressure gas, known as the driver section, whereas another is filled with low-pressure gas, called the driven section. The shock wave initiates when this diaphragm ruptures after the pressure difference across driver and driven side reaches the bursting pressure of the diaphragm. The detailed operation of a shock tube can be found in [4]. Our shock tube is designed in a way that the reflected expansion fan meets the reflected shock before the contact surface, which effectively quenches the reaction. Because of this single pulse functioning, the shock tube is known as a single-pulse shock tube. More details about single pulse shock tube can be obtained in [5]. Figure 5.1 represents the x-t diagram of a shock tube. x indicates the position along the shock tube, and t is the time from the diaphragm ruptures. In our experiments, we use the reflected shock wave to obtain the required operating conditions. As shown in Fig. 5.1, T_5 and P_5 represents the temperature and pressure behind reflected shock wave.

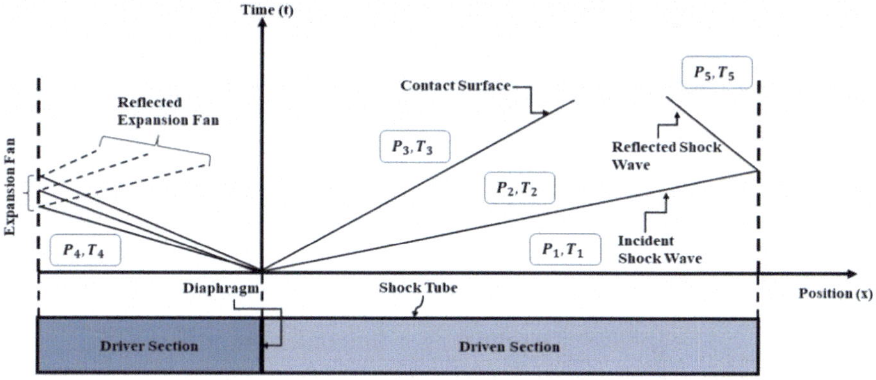

Fig. 5.1 Typical x-t diagram of the shock and expansion wave in a shock tube [4]

5.1.2 Theoretical Temperature Behind Reflected Shock Wave

In a shock tube, the temperature behind the reflected shock wave $(T_{5,theo})$ can be computed using the Rankine-Hugoniot (R-H) relations for normal shock waves. The equation of $T_{5,theo}$ obtained employing the R-H relations can be given as [4]:

$$\frac{T_{5,theo}}{T_1} = \frac{\left[2(\gamma - 1)M^2 + (3 - \gamma)\right]\left[(3\gamma - 1)M^2 - 2(\gamma - 1)\right]}{(\gamma + 1)^2 M^2} \tag{5.1}$$

where T_1 is the initial temperature in the driven section of the shock tube, M is the Mach number of the incident shock wave and γ is the specific heat ratio.

Because of non-ideal effects such as boundary layer formation, real gas effects, and exo/endothermicity of chemical reactions, the temperature calculated using this relation deviates from the actual temperature [4]. This issue can be resolved using a method called chemical thermometry.

5.1.3 Chemical Thermometry

Chemical thermometry uses the kinetics of a known reaction to calibrate the temperature [5]. This article gives a basic overview of the chemical thermometry and the details about it can be found in [6]. Chemical thermometry of a shock tube can be executed using either internal or external standards. Method of using an internal standard involves using a molecule (internal standard) with known kinetic parameters in a shock tube experiment alongside the molecule of interest. Then the progression of internal standard decomposition, reaction dwell time and the rate parameters are utilized to calculate the reaction temperature, i.e., the temperature behind the reflected shock wave $(T_{5,act})$. The following equation is used to determine the $T_{5,act}$

when using an internal standard [7]:

$$T = (-E_a/R)/\left[\ln\left(-\frac{1}{A \times t}\ln(1-\chi)\right)\right] \quad (5.2)$$

where R is the universal gas constant, t is the reaction dwell time, E_a is the activation energy and A is the pre-exponential factor for the decomposition reaction of the internal standard. χ is the extent of internal standard decomposition and it can be given by [7]:

$$\chi = [product]_t/([product]_t + [internalstandard]_t) \quad (5.3)$$

where the $[internal\ standard]_t$ and $[product]_t$ are the concentration of internal standard and decomposition products of internal standard, respectively, at the end of reaction dwell time.

However, there are some limitations with using internal standards for shock tube chemical thermometry. First, the internal standard or its breakdown products should not be identical to any product formed by decomposing molecules of interest. Secondly, neither internal standard nor its products should react either with the compound under investigation or its products. Utilizing the external standard for chemical thermometry could eliminate these limitations. In this method, the experiments with the molecule whose kinetic parameters are already known (external standard) are performed independently of the molecule of interest, covering the temperature range, and the calibration curve is drawn between the measured temperature $(T_{5,act})$ and the temperatures calculated from the Eq. (5.1) $(T_{5,theo})$. For computing the actual temperature behind a reflected shock wave, external standard method also employs Eq. (5.2).

In this study we have presented the chemical thermometry of a single pulse shock tube with cyclohexene as an external standard to measure the actual temperature behind the reflected shock wave $(T_{5,act})$ under low-pressure conditions. This study has led to the development of calibration curve relating the theoretical and actual temperature behind reflected shock wave.

5.2 Experimental Setup

Figure 5.2 shows the schematic of the chemical shock tube (CST-3) present in our laboratory. CST-3 is a cylindrical tube made of stainless steel (grade-304), having an inner and outer diameter of 54 mm and 100 mm respectively. The length of the driver side of the tube is 2.1 m, whereas the length of the driven side is 5.3 m. We used an aluminium diaphragm of 0.8 mm thickness with a groove to separate the driver and driven section. The dump tank is integrated at a 45° angle in the driven part near the diaphragm section. Its primary function is to quench the multiple reflections of a shock wave [8].

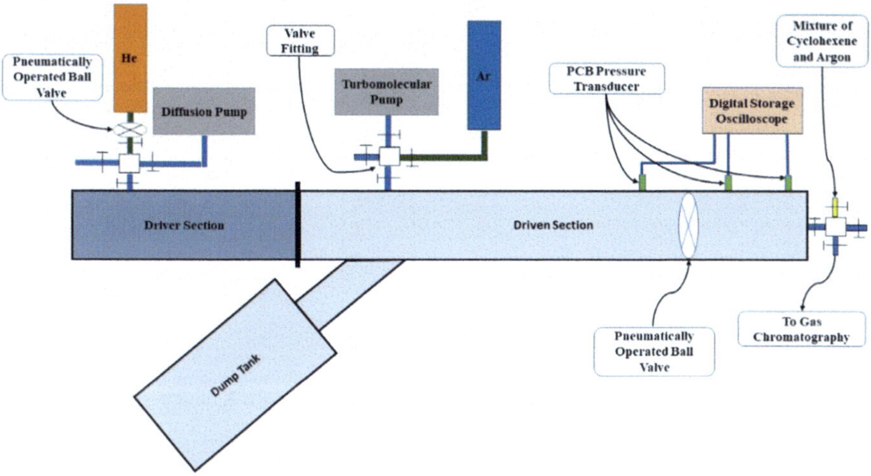

Fig. 5.2 Schematic of chemical shock tube 3 (CST-3)

The driven side of the tube is divided into two sections with the help of a pneumatically operated ball valve. The [9] introduced the ball valve and described its advantages in the chemical kinetics studies, such as achieving uniform test conditions throughout the test sample. The portion of the tube from the ball valve to the end flange is termed as test-section. The implementation of an appropriate piping and fitting successfully facilitated the entry of gases into both sections. This arrangement was also used for evacuating the shock tube before every experiment. A turbomolecular pump (Edward RV12) was used to evacuate the driven side to absolute pressures of 10^8 m bar, while a diffusion pump was used to evacuate the driving side to absolute pressures of 10^2 m bar. During experiments, we filled the driven side excluding test-section with only argon (purity $>$ 99.99%), whereas in the test section, we filled premixed mixture of 1.5% cyclohexene with 98.5% argon. A mixing tube with a 9-L capacity was used to make a mixture of cyclohexene and argon gas. While mixing the pressure of cyclohexene was kept below its vapor pressure to avoid its condensation. In this study, Helium (purity $>$ 99.99%) is used as driver gas. The pressure on the driver and driven sides has been adjusted to obtain the shock wave of different strengths to cover the required temperature range. Driver side pressure can be adjusted using diaphragms with different groove depths. The post-shock mixture was analyzed using a gas chromatograph (GC-7890 B Agilent) with dual FID (flame ionization detector). HP-5 (Cross-linked 5% PH siloxane, 30 m $\times$ 320 μm $\times$ 0.25 μm) and HP-PLOT-Q columns (a bonded polystyrene-divinylbenzene, 30 m $\times$ 320 μm $\times$ 0.20 μm) capillary columns have been utilized in qualitative and quantitative analysis of product species in GC. Nitrogen with purity of 99.999% was employed as a carrier gas, with a flow rate of 0.1 ml/min

Three PCB piezoelectric transducers (Model no. 113B22) are flush mounted at a distance of 5, 450 and 1550 mm from the end flange of the shock tube to monitor

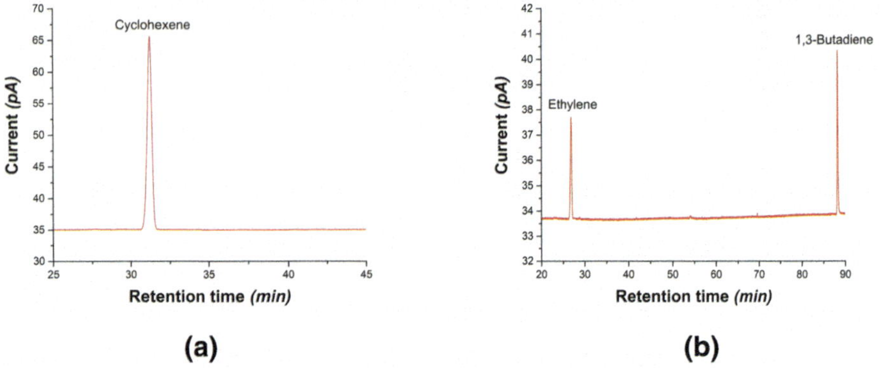

Fig. 5.3 **a** Chromatogram from GC back-FID at $T_{5,act}$ of 1070.81 K and $P_{5,\exp}$ of 3.5 bar, **b** Chromatogram from GC front-FID at $T_{5,act}$ of 1070.81 K and $P_{5,\exp}$ of 3.5 bar

the pressure and to calculate incident shock velocity, which gives the incident shock Mach number. The output of these transducers is connected to the signal conditioner, which in turn is connected to a digital storage oscilloscope (Tektronix, TDS-2014B). The oscilloscope was triggered by the first PCB transducer present in the path of the incident shock wave. The equilibrated pressure in the driven section (P1) and diaphragm bursting pressure in the driver section (P4) was measured by pressure sensors (IRA) with a low acquisition rate. Test cells of volume 1 L are used to collect the sample from the tube and inject it into GC. Typical chromatogram of the reactant and products obtained in the experiments can be seen in the Fig. 3a and b respectively.

5.3 Results and Discussion

The reaction (R1) of thermal decomposition of cyclohexene in ethylene and 1,3-butadiene was used as an external standard as its rate parameters were available in the desired temperature range at low pressures.

$$\text{cyclohexene} \rightarrow \text{ethylene} + 1,3\text{ - butadiene} \tag{R1}$$

Several experiments have been carried out to cover the required temperature range by varying the driver section to driven section pressure ratio (P_4/P_1). These experiments were conducted in the temperature $\left(T_{5,act}\right)$ range of 970–1136 K and pressure $\left(P_{5,act}\right)$ range of 2.2–3.7 bar. Even though the experimental pressure ranged from 2.2 to 3.7 bar, this calibration is valid throughout a pressure range of 0.81–3.75 bar since the rate equation used in this study is pressure independent in that range. For quantification of concentrations, the sensitivity data of cyclohexene, ethylene, and 1,3-butadiene in GC-Flame Ionization Detector (FID) from previous investigations in our lab were used [10, 11].

The rate constant in the experiments (k_{exp}) was determined using the integrated form of rate law for the first order reaction as given below:

$$[C_6H_{10}]_t = [C_6H_{10}]_0 \times e^{-k_{exp}t} \tag{5.4}$$

where $[C_6H_{10}]_0$ and $[C_6H_{10}]_t$ are the initial and final concentration of cyclohexene and t is the reaction dwell time obtained in the experiments.

The experimental temperature $(T_{5,act})$ for the reaction has been calculated by putting the rate constant obtained in experiments (k_{exp}) into the Arrhenius expression reported for the same reaction in the literature [12] for the temperature range of 950–1300 K.

$$k\left[s^{-1}\right] = 4.84 \times 10^{14}\exp(-31900[K]/T) \tag{5.5}$$

In the studied temperature $(T_{5,act})$ range of 970–1136 K, the unimolecular decomposition of cyclohexene (C_6H_{10}) into ethylene (C_2H_4) and 1,3-butadiene (C_4H_6) was the only thermal decomposition pathway. In unimolecular decomposition reaction of hydrocarbon, we can calculate the initial concentration of reactant using two different methods. In first method we determine the initial concentration of reactant using product whose stoichiometric coefficient is fixed with the stoichiometric coefficient of the reactant. In the second method we do the carbon balancing of the product species to determine the initial concentration of reactant. In this study we have calculated the initial concentration of *cyclohexene* $[C_6H_{10}]_0$ using both the methods. Also, we have used the first method twice considering both the products independently. Following are the three path we have used to calculate the initial concentration of *cyclohexene*.

$$Path\,A : [C_6H_{10}]_0 = [C_6H_{10}]_t + [C_2H_4]_t.$$

$$Path\,B : [C_6H_{10}]_0 = [C_6H_{10}]_t + [C_4H_6]_t.$$

$$Path\,C : [C_6H_{10}]_0 = [C_6H_{10}]_t + [C_2H_4]_t/3 + 2[C_4H_6]_t/3.$$

Path A uses the final concentration of cyclohexene and ethylene while the *Path B* use the final concentration of cyclohexene and 1,3-butadiene to calculate the initial concentration of cyclohexene. *Path C* is basically a carbon balancing technique which uses final concentration of reactant and both products. Ideally the initial concentration of cyclohexene calculated from all the three paths must be equivalent. In our study we found slight deviation in initial concentration of cyclohexene obtained from these three paths. This deviation could have been resulted from the uncertainty in the sensitivity data of products and reactant used from previous studies.

Initially we have calculated the T_5 considering the initial concentration of cyclohexene obtained from all the three paths as given in Table 5.1. The maximum deviation in the value of T_5 obtained between any two paths is 23 K. Finally, to minimize the uncertainty, we have taken the average of T_5 obtained from all the three paths, which is considered as the actual temperature ($T_{5,act}$) behind the reflected shock wave.

The calculation of theoretical temperature behind the reflected shock wave $(T_{5,theo})$ requires the value of the incident shock Mach number (M) at the end flange of shock tube. The end flange Mach number (M) of the incident shock wave was calculated from its velocity (U) at the end flange and the speed of sound (a) in the driven section gas. The difference in time between the shock-induced pressure rise and the distance between the pressure transducers was used to calculate the average velocity of the shock wave between pressure transducers. With shock attenuation in consideration, the average velocities obtained with three pressure transducers were extrapolated to the end flange to yield shock velocity at the end flange. During the present study we have observed the incident shock attenuation in the range of 4–7%/m. Substituting the M along with T_1 and γ into Eq. (5.1) gave the value of theoretical temperature $T_{5,theo}$.

Table 5.2 contains data of the $T_{5,theo}$ and $T_{5,act}$ along with percentage deviation ($\%\Delta T$) of $T_{5,theo}$ from $T_{5,act}$ and the corresponding pressure in the experiments.

The least-square fitting of this data has been performed (Fig. 5.4) to determine $T_{5,act}$ with respect to $T_{5,theo}$. The correlation between $T_{5,act}$ and $T_{5,theo}$ within the experimental temperature ($T_{5,act}$) range of 970–1136 K is found to be:

$$T_{5,act} = 0.60914 \times T_{5,theo} + 379.60702\left(R^2 = 0.99015\right) \quad (5.6)$$

It is estimated that the maximum uncertainty in the calibrated temperature is up to 32 K. The Fig. 5.4 clearly shows that the in the studied temperature range,

Table 5.1 Summary of the experimental data of dwell time with the ratio of final to initial concentration of cyclohexene calculated using three paths and corresponding temperature (T_5)

S. No.	Dwell time (ms)	$\left\{\frac{[C_6H_{10}]_t}{[C_6H_{10}]_0}\right\}_A$	$T_{5,A}$(K)	$\left\{\frac{[C_6H_{10}]_t}{[C_6H_{10}]_0}\right\}_B$	$T_{5,B}$(K)	$\left\{\frac{[C_6H_{10}]_t}{[C_6H_{10}]_0}\right\}_C$	$T_{5,C}$(K)
1	1.624	0.99510	975.34	0.99639	966.24	0.99596	969.58
2	1.712	0.98611	1005.91	0.99337	982.84	0.99094	992.45
3	1.568	0.98532	1010.49	0.98877	1001.92	0.98762	1005.03
4	1.972	0.97591	1019.23	0.98512	1003.62	0.98203	1009.66
5	1.608	0.96415	1039.40	0.97693	1024.47	0.97264	1030.19
6	1.592	0.94892	1052.16	0.96209	1041.68	0.95766	1045.53
7	1.512	0.90580	1076.53	0.92881	1066.02	0.92102	1069.88
8	1.664	0.85658	1089.48	0.89073	1078.75	0.87905	1082.71
9	1.856	0.77913	1103.36	0.83124	1092.02	0.81311	1096.24
10	1.788	0.76477	1107.54	0.82512	1094.88	0.80397	1099.66
11	1.768	0.60696	1132.42	0.69842	1119.31	0.66502	1124.35
12	1.684	0.55354	1141.25	0.64034	1129.82	0.60854	1134.17

Table 5.2 $T_{5,act}$ and $T_{5,theo}$ data along with experimental pressure

S. No.	$T_{5,theo}$ (K)	$T_{5,act}$ (K)	$\%\Delta T$	$P_{5,exp}$ (bar)
1	975.48	970.38	0.52	2.8
2	1012.95	993.73	1.89	2.2
3	1024.61	1005.81	1.83	3.2
4	1033.37	1010.84	2.18	3.7
5	1064.27	1031.35	3.09	3.5
6	1104.07	1046.46	5.22	3.1
7	1127.6	1070.81	5.03	3.5
8	1153.83	1083.64	6.08	3.1
9	1167.32	1100.69	5.71	3.7
10	1195.09	1097.21	8.19	3.6
11	1219.82	1125.36	7.74	2.4
12	1245.38	1135.08	8.86	3.5

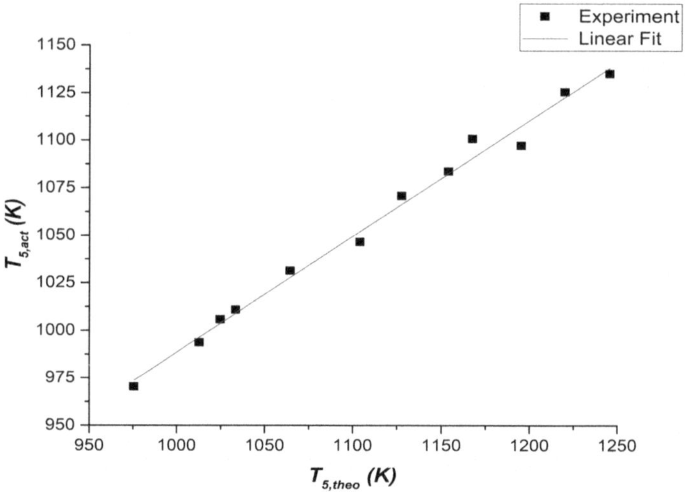

Fig. 5.4 Plot of experimental temperature ($T_{5,act}$) against theoretical temperature ($T_{5,theo}$) in the experimental temperature range of 970–1136 K

$T_{5,theo}$, exceeds the $T_{5,act}$, and the difference grows with increasing temperature. In the calibrated temperature range, the reflected shock temperatures computed using the incident shock Mach number, $T_{5,theo}$, and those obtained by measuring the amount of cyclohexene decomposition, $T_{5,act}$, deviated by 0.52–8.86%. It is also observed that $T_{5,theo}$ estimated by normal shock relations is close to the $T_{5,act}$ predicted by chemical thermometry in the lower temperature range.

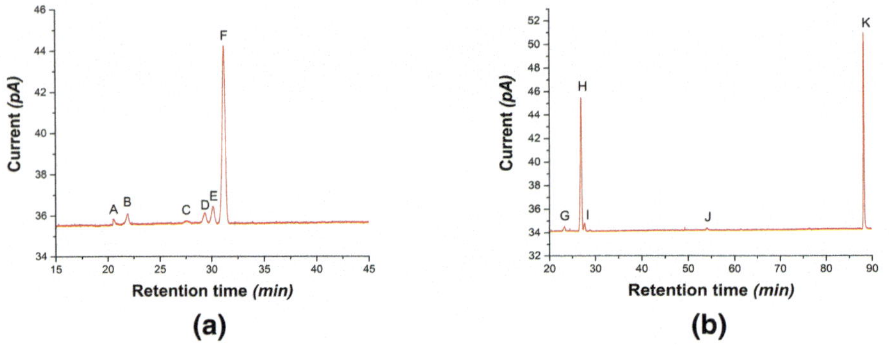

Fig. 5.5 **a** Chromatogram *from GC back-FID at* $T_{5,act}$ *of* 1142.14 K *and* $P_{5,exp}$ *of* 3.4*bar* Peaks are assigned to (A) 1, 3 - pentadiene, (B) cyclopentadiene, (C) methylcyclopentadiene, (D) benzene, (E) cyclohexadiene, (F) cyclohexene, **b** Chromatogram *from GC back-FID at* $T_{5,act}$ *of* 1142.14 K *and* $P_{5,exp}$ *of* 3.4 bar Peaks are assigned to (G) methane, (H) ethylene, (Q) acetylene, (J) propene, (K) 1, 3 - butadiene

Even though [12] suggests that the R1 is the only pathway in the temperature range of 950 to 1300 K for the pressure range of 0.8–3.7 atm, we have observed that the products such as methane, acetylene, propene, 1,3-pentadiene, cyclopentadiene, methylcyclopentadiene, benzene and cyclohexadiene apart from the products mentioned in (R1) at the actual temperature of 1142.14 K (obtained using the Eq. (5.6), corresponding $T_{5,theo}$ of 1251.82 K). Figure 5a and b represent the chromatogram of products of unimolecular decomposition of cyclohexene at 1142.14 K which is lowest temperature where we have observed the products other than ethylene and 1,3-butadiene.

The presence of above products at and above 1142.14 K suggests the strong possibility of secondary reactions as well as secondary pathways of cyclohexene unimolecular decomposition. Therefore, using the given reaction (R1), as a chemical thermometer above 1142.14 K can lead to the wrong prediction of actual temperature due to interference from secondary reaction pathways.

5.4 Conclusion

E temperature calibration of a chemical shock tube (CST-3) has been demonstrated using cyclohexene as an external standard. These experiments were carried out at actual temperatures ranging from 970 to 1136 K. The experimental pressure ranged from 2.2 to 3.7 bar. Although the experimental pressure varied from 2.2 to 3.7 bar, this calibration is valid over a pressure range of 0.81–3.75 bar since the rate equation is pressure independent in that range. The 1-D normal shock relations have been found to overpredict the temperature behind the reflected shock wave, because of the assumptions used in calculations. For the actual temperature ranging from 970 to

1136 K, the theoretical reflected shock temperature overestimated the actual temperature by 0.52–8.86%. This deviation highlights the need of shock tube temperature calibration for obtaining correct kinetic and ignition delay data. Finally, a linear correlation was established between theoretical and actual temperature. We have also discussed the limitation of using cyclohexene as a chemical thermometer above 1142.14 K.

References

1. K.A. Heufer, H. Olivier, Determination of ignition delay times of different hydrocarbons in a new high pressure shock tube. Shock Waves **20**, 307–316 (2010)
2. E.T. Curran, Scramjet engines: the first forty years. J. Propul. Power **17**(6), 1138–1148 (2001)
3. I.W. Kay, W.T. Peschke, R.N. Guile, Hydrocarbon-fueled scramjet combustor investigation. J. Propul. Power **8**(2), 507–512 (1992)
4. A.G. Gaydon, I.R. Hurle, *The Shock Tube in High-Temperature Chemical Physics* (Reinhold Publishing, New York, 1963)
5. W. Tsang, A. Lifshtiz, Shock tube techniques in chemical kinetics. Annu. Rev. Phys. Chem. **41**, 559–599 (1990)
6. K.M. Singh, B. Rajakumar, E. Arunan, Measuring temperature of reflected shock wave using a standard chemical reaction. J. Indian Inst. Sci. **96**, 53–62 (2016)
7. R.S. Tranter, R. Sivaramakrishnan, N. Srinivasan, K. Brezinsky, Calibration of reaction temperatures in a very high pressure shock tube using chemical thermometers. Int. J. Chem. Kinet. **33**(11), 722–731 (2001)
8. A. Lifshitz, S.H. Bauer, E.L. Resler, Studies with a single-pulse shock tube. I. The Cis—trans isomerization of Butene-2. J. Chem. Phys. **38**(10), 2056–2068 (1963)
9. A. Lifshitz, C. Tamburu, A. Suslensky, F. Dubnikova, Decomposition and isomerization of 1, 2-benzisoxazole: single-pulse shock-tube experiments, quantum chemical and transition-state theory calculations. J. Phys. Chem. A **110**, 11677–11683 (2006)
10. S. Hossain, *Shock Tube Experimental and Advanced Computational Investigations on Pyrolysis of Cyclohexane Derivatives and C2 + C2 Reaction* (Doctoral dissertation). Indian Institute of Science, Bangalore, India, (2022)
11. K.V. Dhoke, *Shock Tube Investigation and Modeling of Dicyclopentadiene: Fundamental to Application Reaction* (Doctoral Dissertation). Indian Institute of Science, Bangalore, India (2021)
12. I. Stranic, D.F. Davidson, R.K. Hanson, Shock tube measurements of the rate constant for the reaction cyclohexene → ethylene + 1,3-butadiene. Chem. Phys. Lett. **584**, 18–23 (2013)

Chapter 6
Detached Eddy Simulations of an Incident Shock-Induced Separation

Siva Vayala and R. Sriram

Abstract The numerical investigation of an incident shock wave turbulent boundary layer interaction at $M_\infty = 2.3$ is carried out using Reynolds-averaged Navier-Stokes (RANS) and detached eddy simulations (DES) in a finite difference-based in-house solver. A successfully validated $k - \omega$ RANS model is used to obtain mean flow properties. Further $k - \omega$ model is extended to DES and adaptive DES models to predict the unsteadiness associated with shock motion. For a 2-dimensional grid, it is observed that RANS simulations qualitatively captured all the important mean flow features such as incident shock, separation shock, separation bubble, reflected shock, etc., but overestimated the separation length by 11% compared to existing experimental results and failed to capture the shock unsteadiness. It is also noticed that all kinds of detached eddy simulations produced a continuously enlarging separation bubble for a 2-dimensional grid. An adaptive DES simulation captures wiggles on the outer regions of the boundary layer, which were insignificant in conventional DES. A 3-dimensional adaptive DES simulation (with a span of $0.7\delta_0$) results in shock structure and separation zone similar to 2-dimensional adaptive DES, but showed a vortex break-down in the separation region and stronger wiggles in the outer shear layer region at $t = 2.0\,ms$.

6.1 Introduction

Shock wave turbulent boundary layer interactions (STBLI) in supersonic and hypersonic flows induce flow separation accompanied by an additional shock system with low-frequency, large-scale pulsating motions, which can have detrimental effects on the performance and structural integrity of the high-speed vehicles. In several instances, an interaction of shockwave with the boundary layer instigates flow separation. Often SBLI involves turbulent boundary layer separations, which are associated with large-scale pulsating motions of shock and separated regions. This unsteadiness was found to have a frequency range between $O(100\,Hz)$ to $O(1000\,Hz)$ and

S. Vayala (✉) · R. Sriram
Indian Institute of Technology Madras, Chennai, India
e-mail: ae19d007@smail.iitm.ac.in

© The Author(s), under exclusive license to Springer Nature Singapore Pte Ltd. 2024
G. Rajesh et al. (eds.), *Advances in Shock Interactions*, Springer Proceedings in Physics 408, https://doi.org/10.1007/978-981-97-6099-2_6

termed as low-frequency relative to the frequency of upstream turbulent boundary layer. Erengil and Dolling [1] measured fluctuating pressure values on compression ramp at $M_{\infty} = 5$ and found that the shock motion in the intermittent region has a low frequency of approximately 500 Hz, while the undisturbed upstream boundary layer has a frequency of approximately 50 kHz. Dussauge et al. [2] corroborated the above argument through their experimental studies on impinging shock interaction with boundary layer on a plate and proved the shock frequency is two orders less than the incoming turbulent flow frequency. It is worth studying the unsteady aspects of shock wave turbulent boundary layer interactions and the source of their occurrences owing to their consequences on high-speed vehicles.

It has been believed widely in the research community that there are two potential sources for the shock unsteadiness: One is forcing by upstream turbulent boundary layer, while the other is downstream separated flow influence. In support of the upstream mechanism, Plotkin [22] and Touber and Sandham [8], through their first-order mathematical models, showed that the shock could act as a low pass filter to the turbulence in the upstream undisturbed boundary layer thus causing low-frequency shock motions. Beresh et al. [20] observed correlations between velocity fluctuations and separation shock motions in the flows over a compression ramp at M = 5. Ganapathisubramani et al. [19] argued that the long coherent structures in the upstream turbulent boundary layer could trigger the large-scale motions of the separation line. Despite having several arguments in favor of the upstream mechanism, the downstream mechanism too holds strong possibilities for the shock-motion. Pirozzoli and Grassso [6] from DNS simulations claimed that low-frequency shock motions are resulted due to acoustic resonance from the shear layer above the separation bubble. Dussauge et al. [2] proposed vortical structures in the 3-dimensional separated regions may trigger low-frequency oscillations of the separation shock system. Piponniau et al. [21] argued that the separation bubble undergoes a low-frequency unsteadiness through discharge and recharge mechanisms due to entrainment characteristics of the shear layer. Touber and Sandham [8] and Priebe and Martin [23] stated that an inherent instability in the separation region could be causing these low-frequency motions.

Shock wave boundary layer interactions are found to be challenging to explore through numerical approaches. Marshall and Dolling [3] showed that Reynolds-Averaged Navier Stokes (RANS) turbulence models do not have a mechanism to capture the unsteadiness, thus inaccurate for time-dependent studies of turbulent separated flows. High-fidelity computational methods such as direct numerical simulations (DNS), large-eddy simulations (LES), hybrid RANS-LES simulations, etc., grabbed attention owing to the failure of RANS methods. Pirozzoli and Grasso [6] used DNS methods for impinging shock wave turbulent boundary layer interactions at Mach = 2.25 to characterize the unsteadiness associated with separated flows. Owing to impracticable computational resources and memory required for DNS of high Reynolds number flows, LES came into light where only energetically significant eddies are resolved, while smaller eddies are modeled using sub-grid scale models. Garnier et al. [7] obtained mean flow characteristics on shock-reflection using LES while Touber and Sandham [8, 9] acquired unsteady characteristics for the same case. Despite having reasonable accuracy and computational expense, LES is still found to be difficult for several high Reynolds number flows due to stringent grid requirements in near-wall regions. As a remedy for this, Spalart [10] proposed a hybrid RANS-LES or detached eddy simulation (DES), DES97 model, where attached boundary layer regions are treated with Spalart-Allmaras turbulence model, while the separated regions and eddy regions are treated using LES methods. In earlier studies DES was restricted to S-A model, but had been extended later to various turbulence models like $k-\omega$, $k-\epsilon$, $k-\omega SST$, etc. with the contributions from Shur et al. [11] and Gritskevich et al. [12]. Deck [13], and Roidl et al. [14]) worked on zonal DES methods in which separate sections of the computational domain are treated with RANS or LES explicitly. Spalart et al. [4] and Reddy et al. [16] derived developed versions of DES by mitigating the issues like modeled stress depletion (MSD), log-layer mismatch (LLM), etc., that the original version was suffering form. Since DES works similar to LES in eddying zones, a few researchers, Yin et al. [17], Durbin et al. [18], and Yin and Durbin [5] succeeded to a greater extent in implementing the dynamic nature of the LES method in DES. As a result of these contributions, a new method, 'adaptive DES,' has been developed and proved to be more accurate than the conventional DES methods.

In the present study, a nominally 2-dimensional incident shock-induced separation is investigated numerically using detached eddy simulations (DES). The main focus is given to the implementation of adaptive DES model in the existing finite-difference-based in-house solver. Differences in results for conventional DES and adaptive DES simulations are discussed. The mean flow fields obtained through the preliminary RANS ($k-\omega$ model) simulations are used as an initialization for DES calcualtions. In the present paper, second section provides the overview of numerical parameters considered for the current simulations. Third section deals with the RANS and DES results which are presented in 3.1 and 3.2 respectively. Finally, fourth section discusses the conclusions made from the current work.

6.2 Numerical Methodology

The present works are on the detached eddy simulations of a nominal 2-dimensional shock-induced separation. The computational domain of $25\delta_0$ x $4\delta_0$, (δ_0 is upstream boundary layer thickness) is considered and discretized with the optimum grid (from preliminary RANS simulations) 601 x 151 in streamwise and wall-normal directions respectively. Grids are clustered in the near-wall regions using hyperbolic stretching, and $y^+ \approx 1.4$ is maintained at the wall, while uniform meshing is used for streamwise direction. Compact $4th$ order and Runge-Kutta $4th$ order schemes are used for spatial and temporal discretizations, respectively. $10th$ order implicit filtering, with filtering coefficient, $\alpha_f = 0.495$ is used for numerical stability. CFL number of 0.50 is maintained throughout the computaitons. Fluid (air) is considered to be Newtoniand and fluid viscosity is modelled using Sutherland's law. The adaptive filtering method is employed for shock capturing. The mean inflow profiles are generated

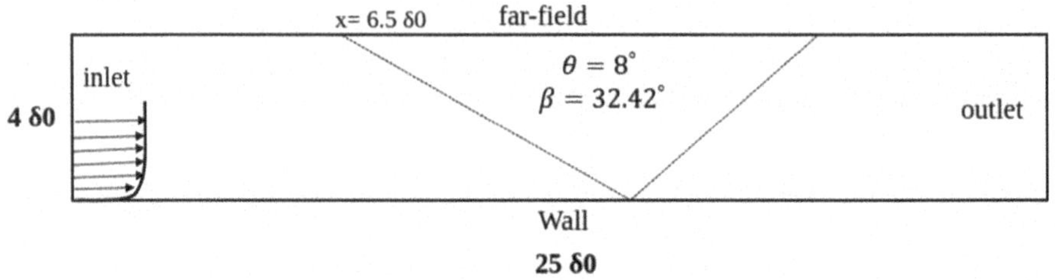

Fig. 6.1 Computational domain

Table 6.1 Flow parameters upstream of the interaction

M_∞	θ	U_∞	δ_0	Re_{δ_0}	C_f	T_{te}
2.3	$8deg$	556 m/s	11 mm	5.82 x 10^4	$\approx$ 2 x 10^{-3}	300 K

using standard wall laws in the van Driest coordinate system, and the shock is generated using Rankine-Hugoniot jump conditions on the top boundary at $x = 6.5\delta_0$. Supersonic inlet condition is provided at the inlet while the outlet is treated with Neumann conditions. Adiabatic and no-slip conditions are applied at the bottom surface, while far-field conditions are applied at the top. The computational domain with dimensions is shown in Fig. 6.1. The flow parameters upstream of the interaction zone are provided in Table 6.1. Comprehensive details of Wilcox's $k - \omega$ turbulence model and adaptive DES models can be found in Wilcox et al. [24] and Yin et al. [5] respectively.

6.3 Results and Discussion

The present solver was made applicable to compressible turbulent flows by incorporating the $k - \omega$ turbulence model and shock-capturing ability using an adaptive filtering approach. Before extending the current solver to DES, preliminary simulations were conducted with the RANS model to assess the mean flow properties and mean shock structure. For this purpose, an impinging shock wave boundary layer interaction was chosen as a test case. The flow parameters are adopted from Dussauge et al. [2]. This case is to corroborate the solver's capabilities prior to extending to 3-dimensional shock-induced separations.

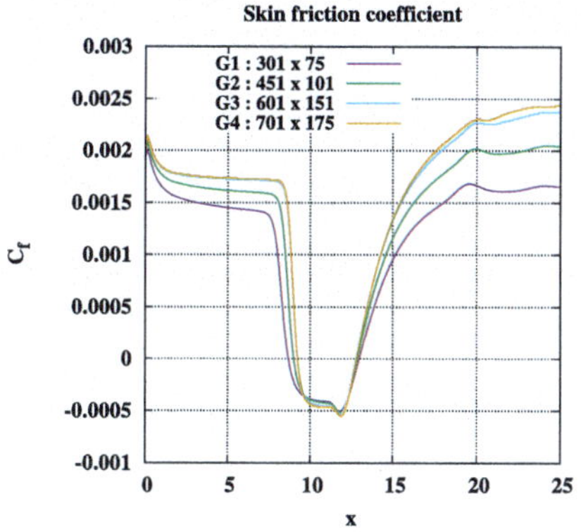

(a) skin friction coefficient

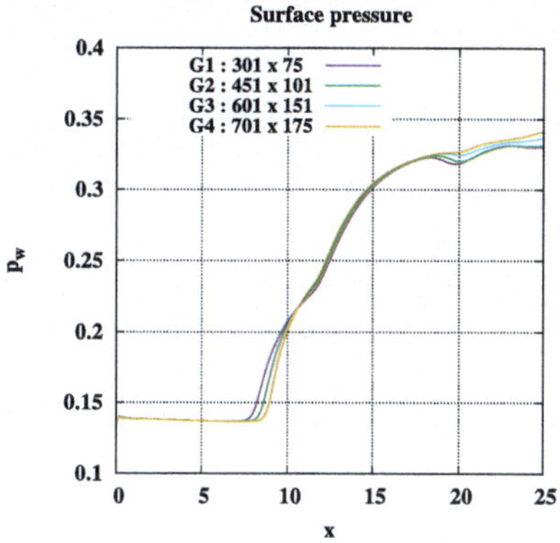

(b) surface pressure

Fig. 6.2 Streamwise distribution of wall properties for different meshes

Table 6.2 Separation details for different grids

Grid	Δx^+	Δy^+	x_s	x_{ra}	x_r	$x_{i,imp}$	L_{sep}	L_{int}
301 × 75	60	2.0	8.69	12.94	7.35	12.67	4.25	5.32
451 × 101	40	1.5	9.01	12.81	7.8	12.67	3.80	4.87
601 × 151	30	1.2	9.25	12.75	8.0	12.67	3.50	4.67
701 × 175	25	1.0	9.25	12.75	8.0	12.67	3.50	4.67

6.3.1 RANS Simulations of 2-Dimensional Incident Shock-Induced Separation

The flow parameters upstream of the interaction zone are given in Table 6.1. Grid sensitivity and time-averaging studies are presented in the following sub sections.
Grid independence study:
For grid sensitivity study, four different grids are considered. In all grids, $y^+ \approx 1$ was maintained near the wall, and grid spacing followed hyperbolic stretching in outer regions, while uniform meshing was applied in the streamwise direction. Turbulent flow was given at the inlet through mean flow profiles using standard wall laws in van Driest coordinate system. All the simulations have been carried out for sufficient time to get a steady shock system. Streamwise distribution of skin friction coefficient and surface pressure for different grids are shown in Fig. 6.2. Separation location, separation length, and interaction length details are presented in the Table 6.2.

Where x_s is separation location, x_{ra} is reattachment location, x_r is inviscid reflected shock location, $x_{i,imp}$ is inviscid shock impingement location, L_{sep} is separation length and L_{int} is the interaction length.

From the Table 6.2, it is noticeable that the grid 601 x 151 is found to be optimum thus used for further studies. Interaction length, $4.67\delta_0$ from the current study is

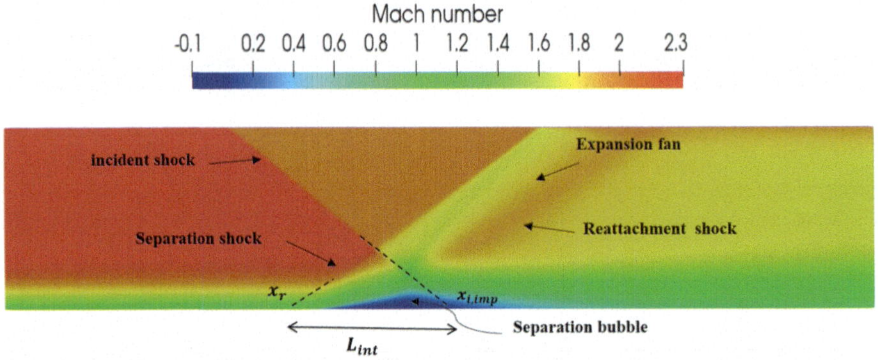

Fig. 6.3 Mach number field for the optimum grid, 601 x 151

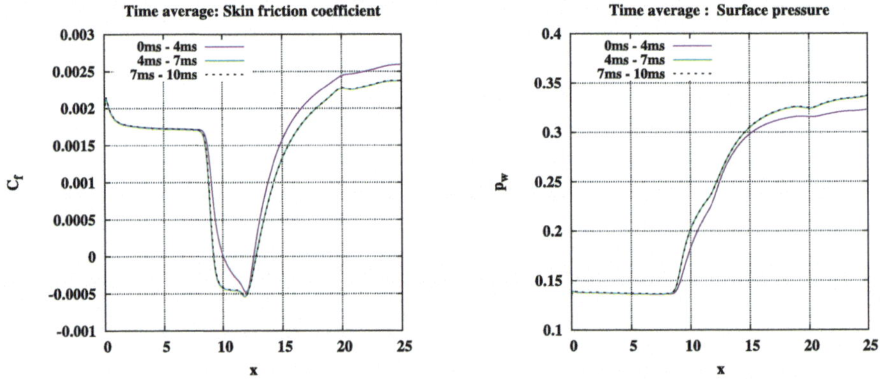

(a) Skin friction coefficient (b) Surface pressure

Fig. 6.4 Time averaged distribution of wall properties

overestimated by $\approx 11\%$ from the experimental value ($\approx 4.2\delta_0$) of Dussauge et al. [2] for the same case. Mach number field for the optimum grid is shown in Fig. 6.3.

Time-average study:
Simulations were carried out for sufficient time, $10ms$, to obtain a steady shock system. From the time-averaged distributions of skin friction coefficient and wall pressure in Fig. 6.4, it is found that at least $4ms$ is required to obtain a steady shock system for RANS simulations, while it is 6ms from the LES studies of Garnier et al. [7].

6.3.2 *Preliminary DES Simulations*

After the successful investigation with the RANS model, the same case is tested with detached eddy simulation (DES). Though turbulence is a 3-dimensional phe-

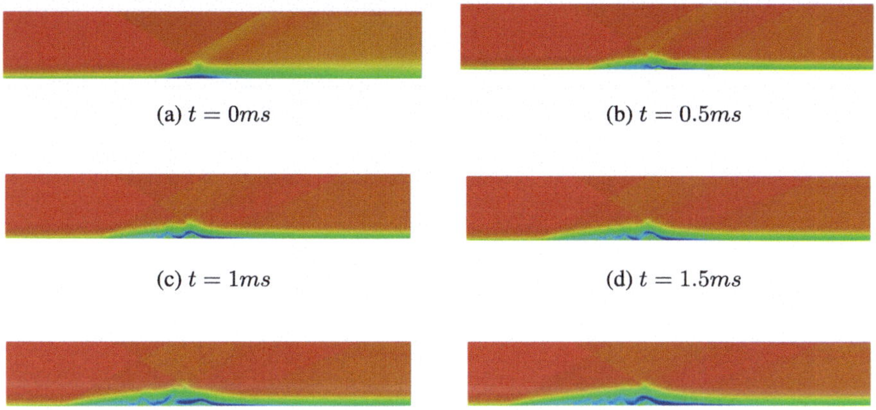

(a) $t = 0ms$ (b) $t = 0.5ms$

(c) $t = 1ms$ (d) $t = 1.5ms$

(e) $t = 2.0ms$ (f) $t = 2.5ms$

Fig. 6.5 Time evolution of interaction using standard DES

nomenon, and DES is not a simulation using averaged equations, the simulation is carried out to know the differences that DES would cause in the solution. An optimum grid from the RANS solution was adopted since it allowed enough grid resolution for LES as well as per Garnier et al. [7], Touber and Sandham [9]. The simulations are initialized with the RANS solution with the same grid and allowed to evolve further. Interestingly, shock and separation bubbles which were expected to undergo oscillations, started moving upstream and growing bigger, respectively like a laminar interaction. The time evolution of the interactions is shown in Fig. 6.5.

The conventional DES method was extended to dynamic DES or adaptive DES by incorporating filtering aspects. Gaussian explicit filtering of Cook and Cabot was used to serve the pupose. From the adpative DES simulation of the same case, wiggles on the outer regions of boundary layer were identified which were barely noticed in conventional DES. This demarcation shows that dynamic nature of DES working . However, it is not appropriate to form conclusion based on 2-dimensional interactions, but difference between DES and adaptive DES results shows a positive indication of working of adaptive DES. Interactions at same instances are shown in Fig. 6.6.

Previous studies shows that 2-dimensional simulations shows a bigger separation bubble size than original 3-dimensional interactions. To check this effect, a 3-dimensional simulation considering a small span-wise width carried-out. Figure 6.7 shows the evolution of the interactions at various time instances. 3-dimensional results seem to be similar to 2-dimensional adaptive DES results, but shown a vortex break-down in the separation region and a stronger wiggles in the outer shear layer region at $t = 2.0\,ms$.

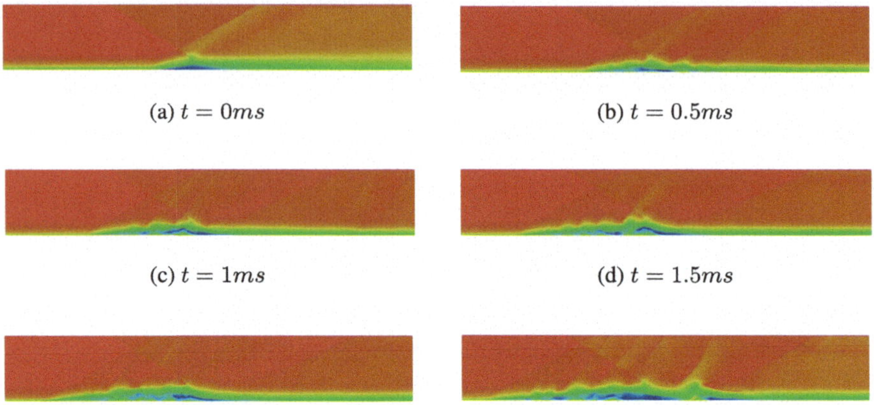

(a) $t = 0ms$ (b) $t = 0.5ms$

(c) $t = 1ms$ (d) $t = 1.5ms$

(e) $t = 2.0ms$ (f) $t = 2.5ms$

Fig. 6.6 Time evolution of interaction using adaptive DES

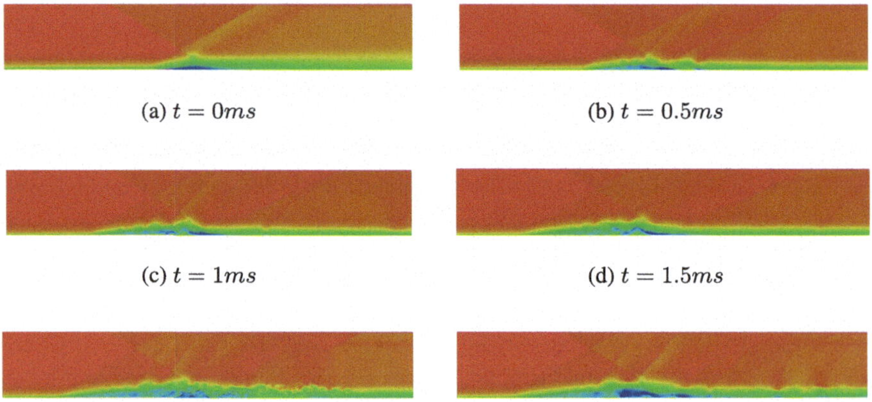

(a) $t = 0ms$ (b) $t = 0.5ms$

(c) $t = 1ms$ (d) $t = 1.5ms$

(e) $t = 2.0ms$ (f) $t = 2.2ms$

Fig. 6.7 Time evolution of interaction using adaptive DES, 3D

6.4 Conclusions

In the present study, an incident shock-induced separation at $M_\infty = 2.3$ is investigated numerically using RANS and DES models in a finite difference-based in-house solver. From RANS simulations, it is found that the separation length has been overestimated by 11% compared to the experimental studies, which is consistent with the literatures. From DES simulations with a 2-dimensional domain, it is noticed that both conventional and adaptive DES models produced enlarging separation bubbles with time. In contrast to conventional DES, an adaptive DES captured wiggles on the outer boundary layer regions, thus showing the signs of the dynamic nature of the model. From 3-dimensional, adaptive DES simulation (with a span of $0.7\delta_0$), a vortex break-down in the separation region and stronger wiggles in the outer shear layer region at $t = 2.0ms$ are observed, despite having similarities with 2-dimensional adaptive DES model in shock structure and separation zone. The enlarging separation zone could be due to the absence of span (in 2-dimensional cases) or insufficient span (in 3-dimensional cases) with appropriate grid resolution. The turbulence production due to vortex break-down in the spanwise direction could have increased the momentum and stabilized the separation bubble to a finite length. The compressible formulations for the adaptive DES model, instead of incompressible formulations, would enhance the predictions further. It would be interesting to explore the potential benefits of the adaptive DES model in capturing the shock unsteadiness with the corrections mentioned above.

References

1. M.E. Erengil, D.S. Dolling, Unsteady wave structure near separation in a mach 5 compression rampinteraction. AIAA J. **29**(5), 728–735 (1991)
2. J.P. Dussauge, P. Dupont, J.F. Debiève, Unsteadiness in shock wave boundary layer interactions with separation. Aerosp. Sci. Technol. **10**(2), 85–91 (2006)
3. T. Marshall, D. Dolling, Computation of turbulent, separated, unswept compression ramp interactions. AIAA J. **30**(8), 2056–2065 (1991)
4. P.R. Spalart, S. Deck, M.L. Shur, K.D. Squires, M.K. Strelets, A. Travin, A new version of detached-eddy simulation, resistant to ambiguous grid densities. Theoret. Comput. Fluid Dyn. **20**(3), 181–195 (2006)
5. Z. Yin, P.A. Durbin, An adaptive des model that allows wall-resolved eddy simulation. Int. J. Heat Fluid Flow **62**, 499–509 (2016)
6. S. Pirozzoli, F. Grasso, Direct numerical simulation of impinging shock wave/turbulent boundary layer interaction at M = 2.25. Phys. Fluids **18**(6), 065–113 (2006)
7. E.P. Garnier, P. Sagaut, M. Deville, Large eddy simulation of shock/boundary-layer interaction. AIAA J. **40**(10), 1935–1944 (2002)
8. E. Touber, N.D. Sandham, Large-eddy simulation of low-frequency unsteadiness in a turbulent shock-induced separation bubble. Theoret. Comput. Fluid Dyn. **23**(2), 79–107 (2009)
9. E. Touber, N.D. Sandham, Low-order stochastic modelling of low-frequency motions in reflected shock-wave/boundary-layer interactions. J. Fluid Mech. **671**, 417–465 (2011)
10. P. R. Spalart, Comments on the feasibility of les for wings, and on a hybrid rans/les approach. In: Proceedings of first AFOSR international conference on DNS/LES. Greyden Press (1997)
11. M.L. Shur, R.R. Spalart, M.K. Strelets, A.K. Travin, A hybrid rans-les approach with delayed-des and wall-modelled les capabilities. Int. J. Heat Fluid Flow **29**(6), 1638–1649 (2008)
12. M.S. Gritskevich, A.V. Garbaruk, J. Schütze, F.R. Menter, Development of DDES and IDDES formulations for the $k - \omega$ shear stress transport model. Flow Turbul. Combust. **88**(3), 431–449 (2012)
13. S. Deck, Recent improvements in the zonal detached eddy simulation (zdes) formulation. Theoret. Comput. Fluid Dyn. **26**(6), 523–550 (2012)
14. B. Roidl, M. Meinke, W. Schröder, A zonal rans-les method for compressible flows. Comput. Fluids **67**, 1–15 (2012)
15. C. Mockett, A comprehensive study of detached eddy simulation. Univerlagtuberlin (2009)
16. K. Reddy, J. Ryon, P. Durbin, A DDES model with a smagorinsky-type eddy viscosity formulation and log-layer mismatch correction. Int. J. Heat Fluid Flow **50**, 103–113 (2014)

17. Z. Yin, K. Reddy, P.A. Durbin, On the dynamic computation of the model constant in delayed detached eddy simulation. Phys. Fluids **27**(2), 025105 (2015)
18. P. Durbin, Z. Yin, E. Jeyapaul, Adaptive detached-eddy simulation of three-dimensional diffusers. J. Fluids Eng. **138**(10), (2016)
19. B. Ganapathisubramani, N. Clemens, D. Dolling, Effects of upstream boundary layer on the unsteadiness of shock-induced separation. J. Fluid Mech. **585**, 369–394 (2007)
20. S. Beresh, N. Clemens, D. Dolling, Relationship between upstream turbulent boundary-layer velocity fluctuations and separation shock unsteadiness. AIAA J. **40**(12), 2412–2422 (2002)
21. S. Piponniau, J.-P. Dussauge, J.-F. Debieve, P. Dupont, A simple model for low-frequency unsteadiness in shock-induced separation. J. Fluid Mech. **629**, 87–108 (2009)
22. K.J. Plotkin, Shock wave oscillation driven by turbulent boundary-layer fluctuations. AIAA J. **13**(8), 1036–1040 (1975)
23. S. Priebe, M.P. Martín, Low-frequency unsteadiness in shock wave-turbulent boundary layer interaction. J. Fluid Mech. **699**, 1–49 (2012)
24. D.C. Wilcox et al., *Turbulence modeling for CFD*, vol. 2 (DCW industries La Canada, CA, 1998)

Chapter 7
Blast Wave Mitigation in Novel Bio-inspired Inhomogeneous Structures

Manish Suresh Raut, Harshit Verma, Mital Patel, S. Gopalakrishnan, and Gopalan Jagadeesh

Abstract Engineering structures are prone to blast loads, especially in aerospace, marine and nuclear sectors. Due to the serious need of sufficiently mitigating these loads, this topic is being heavily researched. Interestingly, it is observed that biological structures are inherently designed to deal with extreme load conditions, for example, the beak of a woodpecker. Thus, looking for the solution in nature, wave propagation in various bio-inspired structures have been studied in the recent past, which holds a vast promise for its application in blast mitigation. Of these, various models including sandwich structures, with sinusoidal depth variations, are discussed and investigated in this study. Firstly, finite element modeling is done with the same blast loads and boundary conditions and other viabilities of the shock experiments to obtain the responses. In order to measure the actual response of the structures to blast loadings, a sample is fabricated, experimented and studied in a vertical shock tube facility. Experimental results are also used to validate the finite element analysis results. The study of the propagation of blast wave in these inhomogeneous media having time-dependent material properties in addition to their interaction with the boundaries aids to understand the energy dissipation of the wave fronts. More results for these models are obtained, which gives us more insight into its critical wave propagation characteristics, and for the purpose of overall comparison of the blast mitigation performance. Blast loads of different amplitudes and decay times are considered in order to understand the blast mitigation characteristics of these structures better.

7.1 Introduction

Blast waves cause very heavy, and mostly irrecoverable, damage to the structures and entities it interacts with. Resistance and early attenuation of the blast like loadings, which can be caused due to, for example, blast or ballistic loads, is the major focus in the development of defence applications including military vehicles like aircraft and ships. Various structural systems have been developed to enhance the blast attenua-

M. S. Raut (✉) · H. Verma · M. Patel · S. Gopalakrishnan · G. Jagadeesh
Department of Aerospace Engineering, Indian Institute of Science, Bangalore 560012, India
e-mail: manishraut@iisc.ac.in

© The Author(s), under exclusive license to Springer Nature Singapore Pte Ltd. 2024

G. Rajesh et al. (eds.), *Advances in Shock Interactions*, Springer Proceedings in Physics 408, https://doi.org/10.1007/978-981-97-6099-2_7

tion characteristics, for example, sandwich plates with sufficiently strong cores have the potential to sustain substantially larger shock loads than monolithic plates of the same material and weight [1]. Core materials of different types have been studied, for example, fluid impregnated foam [2] were investigated to understand their potential to absorb energy and impede shock wave. Also foam cores with filler materials like glass beads and tuff volcanic rock [3] were studied to see their effect on the shock responses and their mode of shock mitigation, either reflective or absorptive. Also, Nonlinear granular matter, which has rich wave dynamics, has high potential for mitigating shocks. The phenomena of energy trapping and shock disintegration is observed in composite granular media consisting of alternating ensembles of beads with high and low elastic moduli of different masses [4]. After the rigorous research in the conventional structural domain, novel structures are being investigated for blast resistance enhancement and refinement. Interestingly, it is observed that biological structures are inherently designed to deal with extreme load conditions. Thus, looking for the solution in nature, various bio-inspired structures have been studied from the viewpoint of blast wave mitigation. Nacre-like composite plates [5] and Bouligand inspired bionic structures [6] were numerically investigated for dynamic loading mitigation performance, and were proposed to weaken the impulse transmission effectively. An another intriguing creature woodpecker remains unharmed without any Traumatic Brain Injury (TBI) despite facing g-forces above 1000g, repeatedly, a few thousand times daily. It is found that the outer layer of their beak has peculiar adaptive wavy sutures between the keratin scales, unlike most of the other birds. The wavy suture interfaces, moreover, are found in various other biological tissues demanding resistance to impact. Recently, longitudinal [7] and flexural [8] wave propagation studies, both elastic and viscoelastic, are thoroughly carried out in waveguides with various sinusoidally varying depth profiles inspired from the suture configurations (Fig. 7.1).

Formulating special superconvergent finite elements for such structures, the characteristic wave propagation behaviour of these waveguides is ascertained there. The *Convex* type waveguides showed amplitude reduction, the *Concave* type waveguides reduced group speeds with increase in the wave amplitude, whereas the *Full* type waveguide reduced group speeds significantly with comparatively less increase in the wave amplitude. For completeness, the depth variation $h(x)$ and area $A(x)$ of a representative segment of the *Convex* waveguide [8] are given below.

$$h(x) = h_0(1 + \alpha \sin\beta x), \ \ A(x) = A_0(1 + \alpha \sin\beta x), \ \ 0 < x < p \tag{7.1}$$

where α and β are the constants controlling the undulation amplitude and the width of the segment. Similarly, the *Concave* waveguide is characterized by replacing the '+' sign in Eq. (7.1) by '–', and the *Full* waveguide is specified by changing the signs alternately that forms a full sinusoidal function. The waveguide with $\alpha = 0$ refers to a uniform or straight waveguide. Based on the numerical studies carried out there, certain promising waveguides which reduce both the group speeds as well as the wave amplitudes are proposed.

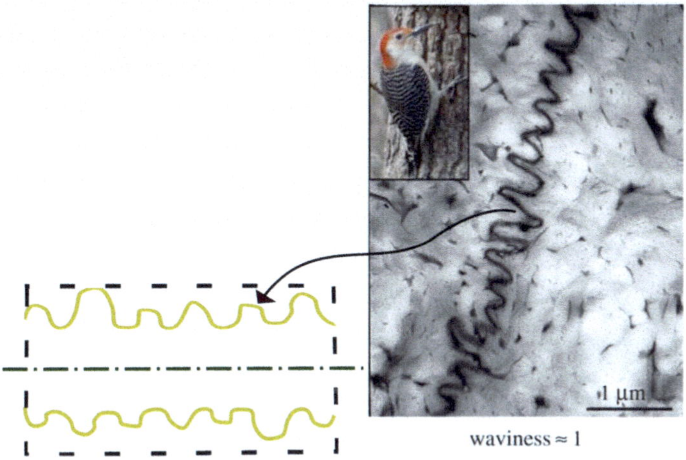

Fig. 7.1 Suture profile in the woodpecker beak and a schematic diagram of the suture structure

Building on the proposed and further optimised configurations, these models, including sandwich structures made of dissipative core and stiff supporting structures, can be studied for shock loadings. In this study, mainly, the model with $5X1C5X$ [8] (the core in Fig. 7.2a) is investigated in detail. The sandwich model with $5X1C5X$ waveguide configuration is fabricated, experimented and studied in shock tubes. The resulting accelerations are then compared with the finite element results. More results for these models are obtained which gives us more insight into the wave characteristics and, for overall comparison with straight and homogeneous waveguides for blast mitigation performance. For better investigation, the results are also obtained for different blast loads. The study of blast wave propagation in these inhomogeneous media having time-dependent material properties along with the interaction of the waves with their boundaries aids to understand the energy dissipation of the wave fronts. The objective of the study is to obtain and analyse both qualitative and quantitative response data for various such assemblies composed of elastic and viscoelastic material components subjected to shock loads, the output of which can then be implemented for shock wave mitigation in the sandwich structures used in aircrafts and marine vessels.

7.2 Description of the Model

The effect of the sinusoidal depth variations on wave propagation, though, is observed by considering linear elastic and the Kelvin-Voigt viscoelastic constitutive models (in flexural formulation) for numerical investigation, sutures as seen in the woodpecker beak have significant collagen content [9]. For the purpose of full analysis of such structures, in this paper, the waveguides made of dissipative material are

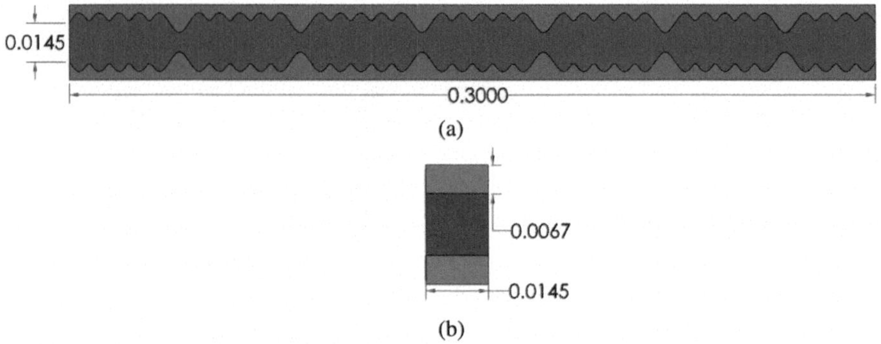

Fig. 7.2 **a** The sandwich model having core waveguide with $n = 40$ and $5X1C5X$ configuration having $\alpha = \pm 0.5$ **b** Cross sectional area at the boundaries of the model (all dimensions are in meters)

considered. The material used is the synthetic Chloroprene Rubber (CR) that is also known by the trade name Neoprene. Neoprene exhibits good chemical stability and relative inertness to degradation in wide varieties of environments. Also it has good mechanical properties including good flexibility over a wide range of temperature which suits for the experimental purposes.

A column of an elastomer buckles in compression at low critical loads due to a very low value of its relaxation modulus, both instantaneous and long term, though this buckling is reversible unlike in metals. To arrest this buckling phenomenon, the waveguides as shown in Fig. 7.2 are sandwiched between two stiff plates. Also, in the woodpecker beak, the suture structures are surrounded by a hard material having larger mineral content, this is an another reason that the rubber core is sandwiched between two stiff materials in this study. Considering the dispersive behaviour of the inhomogeneous periodic waveguides and the time dependent dissipative nature of the material used, studying outcome of their coupling on the mitigation characteristics is interesting. Figure 7.2 shows the dimensions of the model used. The length of the sandwich structure is 0.3 m, with total number of segments $n = 40$, and the end cross section of the core is 14.5 mm $\times$ 14.5 mm and that for each aluminium plate is 14.5 mm $\times$ 6.7 mm. Keeping these dimensions same in all the models, the height of the core profile can be varied with different values of α.

7.3 Methods and Investigation

7.3.1 Finite Element Analysis

Finite element (FE) analysis comprises of discretization of the domain, interpolation of the field variables and geometry, implementation of the material and strain-

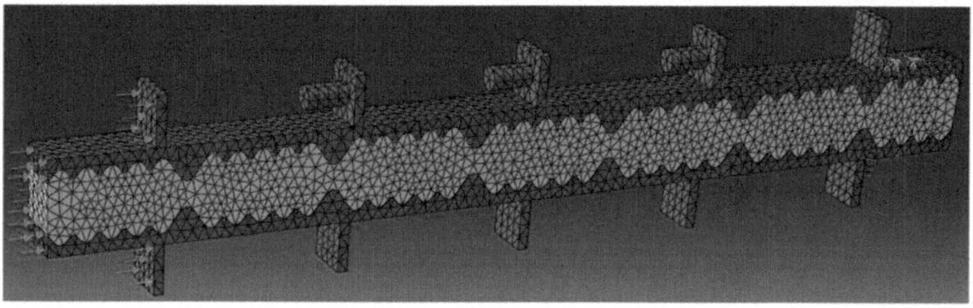

Fig. 7.3 FE model with mesh and boundary conditions

displacement models and the boundary conditions before solving the corresponding differential equations for given loads. In this study, commercial FE code SIMULIA Abaqus is used to do the analysis.

The sandwich model shown in Fig. 7.2a is first considered for analysis due to its easiness for installing sensors in physical experiments. A slight modification of the sandwich structure is done to incorporate the installation of the sensors and these modifications are considered in the FE analysis as can be seen in Fig. 7.3. The domain is discretized into 32205 10-noded quadratic tetrahedron elements.

Since rubber-like elastomers exhibit large deformations with nonlinear stress-strain curve that is attributed to the nonlinear deformation of the cross-linked polymer network, along with dissipation of energy due to the diffusion of free chains [10], they are modeled using finite strain viscoelastic constitutive models. This is usually accomplished by considering a hyperelastic model in conjunction with a viscoelastic model. The hyperelastic material models are non-linear which shows elastic response for large strains, and hence need to also account for geometric nonlinearity while solving. In addition to often being incompressible or very nearly so, the hyperelastic materials are assumed isotropic throughout the deformation history, which allows to conveniently formulate strain energy potentials as a function of the strain invariants. This description fits well for elastic behaviour of the rubber-like elastomers. There exist hyperelastic models with different strain energy potentials, which define the strain energy stored in the material per unit reference volume as a function of strain at that point in the material. In this study, Arruda-Boyce hyperelastic model which is a mechanistic model is used. The strain energy potential w for the Arruda-Boyce model has the following form.

$$w = \mu \sum_{i=1}^{5} \frac{C_i}{\lambda_m^{2i-2}} \left(\overline{I}_1^{\,i} - 3^i \right) + \frac{1}{D} \left(\frac{J_{el}^2 - 1}{2} - \ln J_{el} \right) \tag{7.2}$$

where μ, λ_m, D are the material parameters related to initial shear modulus, locking stretch and initial bulk modulus, respectively, $\overline{I}_1$ is the first deviatoric strain invariant which is a function of the principle stretches, J_{el} is the elastic volume ratio and the sequence

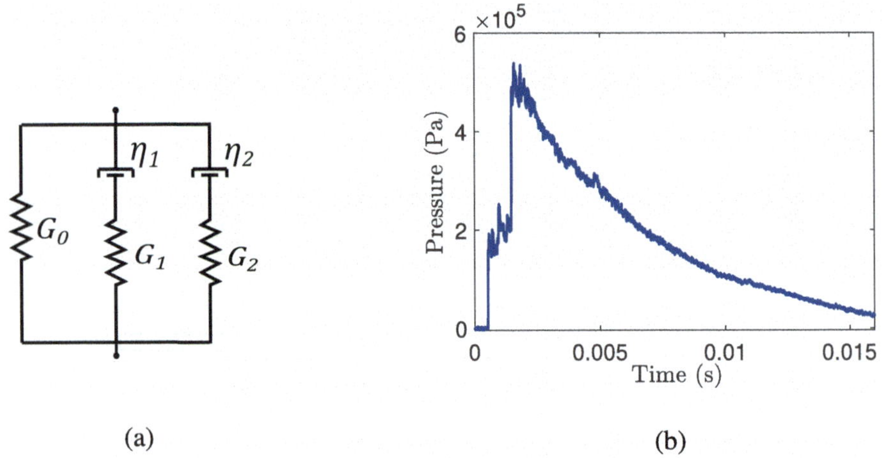

Fig. 7.4 **a** Generalized Maxwell model for viscoelastic modeling **b** Blast load profile

$$(C_i) := \left(\frac{1}{2}, \frac{1}{20}, \frac{11}{1050}, \frac{19}{7000}, \frac{519}{673750}\right) \tag{7.3}$$

The Generalised Maxwell model as shown in Fig. 7.4a, which has two Maxwell elements in parallel, is used for modeling viscoelasticity in this study. The associated three term Prony series that represents the normalized shear relaxation modulus $g(t)$, with initial shear modulus as G_0, is given by

$$g(t) = 1 - \sum_{i=1}^{2} g_i \left(1 - e^{-\frac{t}{\tau_i}}\right) \tag{7.4}$$

$$\text{where} \quad g_i = \frac{G_i}{G_0}, \quad \text{and relaxation time} \quad \tau_i = \frac{\eta_i}{G_0} \tag{7.5}$$

Similar expression is used to represent the bulk modulus, assuming the isotropic material properties. Assuming incompressibility, the material parameters for the hyperelastic and viscoelastic models are obtained experimentally from various tests on neoprene rubber with 50 Shore A, respectively, as $\mu = 1.971042$ MPa, $\lambda_m = 3.95596$, $g_1 = 0.12548$, $g_2 = 0.056483$, $\tau_1 = 0.00751722$ and $\tau_2 = 460.64$, with its density as 1230 kg m^{-3}. The material of the supporting structure is considered to be aluminium 6061. The experimentally obtained blast load data as shown in Fig. 7.4b is used to give axial pressure load in the FE model, and Dirichlet boundary conditions are imposed on few nodes at one end of the structure as shown in Fig. 7.3. An explicit time integration scheme is used to obtain the responses (Fig. 7.5).

Fig. 7.5 Vertical Shock Tube facility in LHSR, IISc Bangalore used for experiments

7.3.2 Experimental Analysis and Discussion

The objective of the experimental study is to obtain responses of the sandwich structures at various locations along its length as it is subjected at one end with a shock load axially in order to compare them with the FE model. Multiple thin symmetric protrusion fins needed to be attached to the aluminium structure for supporting the three uniaxial PCB® accelerometers which were mounted, at equal distances, as shown in the Fig. 7.6. An ethanol based adhesive is used for bonding the rubber and aluminium parts. The experiments are performed in the Vertical Shock Tube (VST) (Fig. 7.5) in Laboratory for Hypersonic and Shockwave Research (LHSR), IISc. The pressure measured using a pressure sensor at the free end of the sandwich structure is same as shown in Fig. 7.4b.

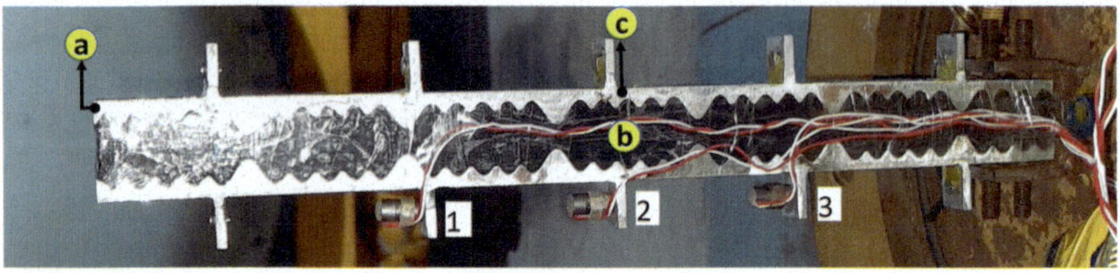

Fig. 7.6 The sandwich specimen (in horizontal orientation) tested in VST. Accelerometers are attached at the three locations, 1, 2 and 3, as shown

The accelerations obtained at two different locations are compared between the experiment and the FE analysis in Fig. 7.7. The incident pressure pulse with (values reduced by 100) is also superposed to set the mark. It can be seen in the figure that except the initial peak, the magnitude of accelerations obtained from FE model at later times is matching with those from the experiment, though the phases are off a bit at some time instants. The peak accelerations as seen from the FE model is going above $500g$. Since the range of the accelerometers used for the experiment is $\pm 450g$, the authors understand that the data given by the accelerometers at those time instant is spurious. Also, as can be seen at further time instants, where the acceleration is sufficiently within the range $\pm 450g$, the values are matching. An exact match is not expected owing to the inexact contacts and joints and the complex constitutive and strain-displacement relations in the experiment. Nevertheless, in the zoomed image as shown in the figures, it can be seen that the responses at the location 3 are appearing later in time than those at location 1, and qualitatively the results are matching closely. Since the wave in aluminium structure will be traveling with speed around 5000 ms^{-1}, the wave will have multiple reflections within hundreds of microseconds as visible in the image.

For the purpose of assessing the performance of these structures, the stress histories for the sandwich structure with $5X1C5X$ neoprene core (sut_{cr}) are compared with the sandwich structures whose cores are straight and are purely made of aluminium (str_{al}) and neoprene (str_{cr}) as well. Also, a blast profile with higher peak pressure of 7 bar is used for these studies for better understanding at different pressures. As can be seen from Fig. 7.8a, at location a in the aluminium cover, which is the impact location, the stress amplitude is very low for the case of the sut_{cr} compared to that of str_{cr}, which is lower than that of str_{al}. At the location b, which is at half the length in the core, the stresses, as shown in Fig. 7.8b, are almost diminishing in sut_{cr} and str_{cr}, while in the case of str_{al}, the stress level is not reducing. When the stress history is zoomed about the time of arrival of the stress wave in the core (Fig. 7.8c), it can be observed that the stress wave in the case of sut_{cr} appears more than 100 μs later than str_{cr}. Similarly, at location c, which is half the length in the aluminium cover, the stress wave in the case of sut_{cr} appear later than both str_{cr} and str_{al} as can be seen in Fig. 7.8d. This phenomena of lower energy speeds in the sut_{cr} type of structures can significantly be exploited to protect critical entities from

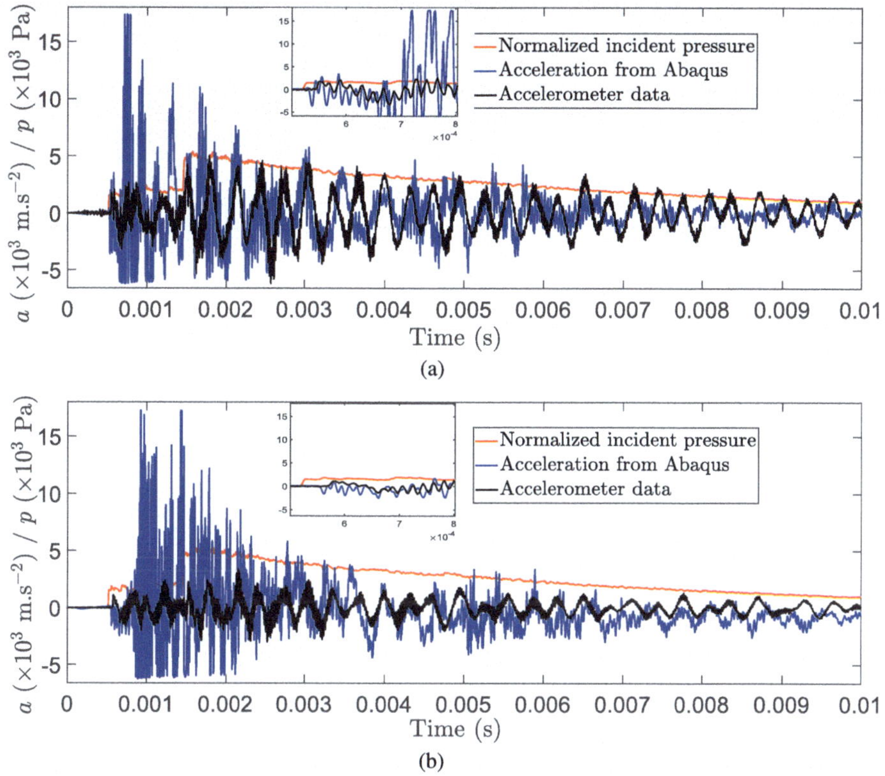

Fig. 7.7 Comparison of the accelerations between experiment and the FE method, also zoomed for initial 0.3 ms **a** at the location 1 and **b** at the location 3

high frequency blast loads having total time (rise time and decay time) of the order of microseconds. Also, additionally the stresses in the CR core almost completely diminishes due to its dissipative nature as the wave propagates along the length.

7.4 Conclusion

In this study, blast wave mitigation characteristics of the novel waveguides inspired from the suture structures observed in nature, in the woodpecker beak in particular, is investigated. Considering the constituent material as neoprene for the $5X1C5X$ configuration waveguide and the aluminium supports for stiffening the structure, their responses to the different blast loads are assessed both experimentally and numerically using the FE analysis. The results from the experiment performed in a vertical shock tube are comparable with the FE results. The stress histories obtained from the finite element studies considering a finite strain viscoelastic model for

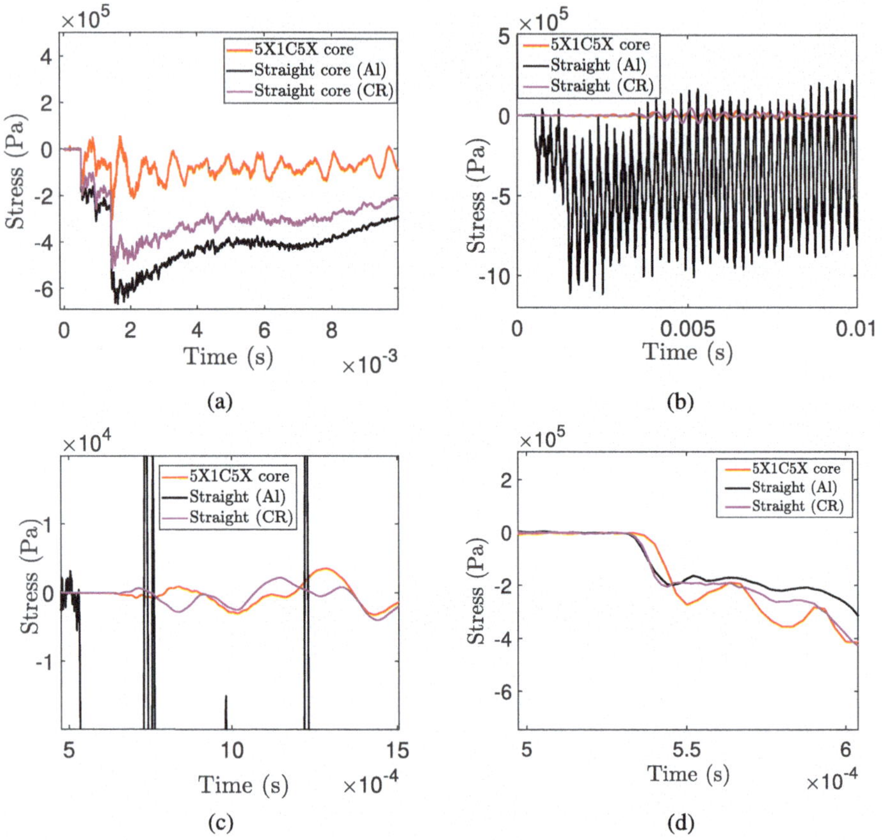

Fig. 7.8 Stress history at **a** location a (from Fig. 7.6) **b** location b, zoomed stress history at **c** location b **d** location c

neoprene, suggest that along with the stress reduction, the energy propagation gets delayed in these structures which is observed of the order of microseconds. A further detailed research needs to be carried out to obtain more time delay as per application needs. Additional examination of the results and improvement of the constitutive models and the experimental setup in terms of better contacts, joints and boundary conditions in order to get the better experimental results are under process. In future study, the attenuation characteristics of these structures for transverse blast loads and waveguide configurations obtained from a detailed multi-objective optimization will be investigated for better blast mitigation functionalities.

Acknowledgements The authors wish to thank Dr. Sachchidanand Das, Scientist F, DMDE, Hyderabad for providing the test data for neoprene, and also the Office of Naval Research (ONR, USA) Solid Mechanics Program for its support.

References

1. Z. Xue, J.W. Hutchinson, Preliminary assessment of sandwich plates subject to blast loads. Int. J. Mech. Sci. **45**(4), 687–705 (2003)
2. M.A. Dawson, G.H. McKinley, L.J. Gibson, The dynamic compressive response of open-cell foam impregnated with a Newtonian fluid. J. Appl. Mech. **75**(4), 041015 (2008)
3. B. Schimizze et al., An experimental and numerical study of blast induced shock wave mitigation in sandwich structures. Appl. Acoust. **74**(1), 1–9 (2013)
4. C. Daraio et al., Energy trapping and shock disintegration in a composite granular medium. Phys. Rev. Lett. **6**(5), 058002 (2006)
5. E.A. Flores-Johnson et al., A numerical study of bioinspired nacre-like composite plates under blast loading. Compos. Struct. **126**, 329–336 (2015)
6. F. Yang, W. Xie, S. Meng, Impact and blast performance enhancement in bio-inspired helicoidal structures: a numerical study. J. Mech. Phys. Solids **142**, 104025 (2020)
7. S. Gopalakrishnan, M.S. Raut, Longitudinal wave propagation in one-dimensional waveguides with sinusoidally varying depth. J. Sound Vib. **463**, 114945 (2019)
8. M.S. Raut, S. Gopalakrishnan, Elastic and viscoelastic flexural wave motion in woodpecker-beak-inspired structures. Bioinspiration & Biomimetics **16**, 046021 (2021)
9. N. Lee et al., Hierarchical multiscale structure-property relationships of the red-bellied woodpecker (Melanerpes carolinus) beak. J. R. Soc. Interface **11**(96), 20140274 (2014)
10. Ying Li et al., Molecular simulation guided constitutive modeling on finite strain viscoelasticity of elastomers. J. Mech. Phys. Solids **88**, 204–226 (2016)

Chapter 8
Effect of Polynomial Degree on Discontinuous Galerkin Simulation of Euler Equations

Vachan D. Potluri, Bhalchandra P. Puranik, and Kowsik V. R. Bodi

Abstract Second order finite volume algorithms have long been used for robust simulation of compressible flows. Discontinuous Galerkin methods constitute one part of a suite of algorithms being developed for high order accurate simulations of such flows. These methods rely on a concept called limiter for stable computation of discontinuous flows. In general, these limiters introduce additional dissipation near discontinuities by locally compromising on the high order accuracy to some extent. For assessing the benefit of a high order simulation employing such a limiter, a shock tube problem and two shock interaction problems are simulated at different resolutions, using several polynomial degrees. Results obtained in these cases show that in flows with smooth variation and/or weak discontinuities, higher order simulations give more accurate results, and this improvement is more marked at lower resolutions. In flows having strong discontinuities, the performance improvement in using higher than quadratic basis is minimal. The relative cost is similar for simulations using quadratic and higher basis.

8.1 Introduction

Development of robust algorithms for numerical simulation of compressible flows has been in active pursuit since the last five decades. Initial efforts on obtaining solutions which are oscillation-free near discontinuities and second order accurate in smooth regions, led to the development of several well-known numerical flux schemes and reconstruction techniques. Focus was then shifted to obtain high order accurate solutions, while simultaneously retaining non-oscillatory nature near discontinuities. One set of methods which are being developed in this line are the discontinuous Galerkin (DG) methods. They have garnered significant interest in the past two decades since they (i) can perform arbitrarily high order accurate computations without much increase in algorithmic complexity, (ii) conveniently handle complex and/or curved meshes, and (iii) have high suitability to hp-adaptivity and

V. D. Potluri (✉) · B. P. Puranik · K. V. R. Bodi
Indian Institute of Technology Bombay, Mumbai, Maharashtra 400076, India
e-mail: vachanpotluri@iitb.ac.in

© The Author(s), under exclusive license to Springer Nature Singapore Pte Ltd. 2024

G. Rajesh et al. (eds.), *Advances in Shock Interactions*, Springer Proceedings in Physics 408, https://doi.org/10.1007/978-981-97-6099-2_8

parallelisation owing to compact cell-local interpolation [1]. In particular, the compact polynomial basis used in these methods readily allows the use of the available robust numerical flux schemes at the element interfaces, leading to a stable discretisation of the convection term [2]. However, this does not guarantee a stable resolution of the discontinuity, since the differential equation itself is invalid in its vicinity [1]. For stable computation of discontinuous solutions the use of a "limiter" is required. Recently, subcell limiters are being developed which split a finite element near a discontinuity into a specific number of subcells, and use a finite volume method (FVM)-inspired discretisation, in place of a pure DG discretisation. The idea is to rely on the robust second order FVM algorithms for stable computation in the elements near discontinuities. Given this nature of a subcell limiter, it is useful to investigate the benefit in performing a DG simulation using high order polynomial basis against a second order solution, in flows dominated by discontinuities.

In this work, the effect of polynomial degree on DG simulations of a set of shock interaction problems is studied using a recently developed subcell limiter [3]. In Sect. 8.3, a simple shock tube problem (modified Sod test), and two shock interaction problems (Shu-Osher test in one-dimension and double Mach reflection in two-dimensions) are considered for this study. These three test cases are chosen to represent three different kinds of flow, with minimal similarity between them. A detailed analysis of the performance of different polynomial degrees is presented for each of these cases. The key findings are summarised, and the article is concluded in Sect. 8.4.

8.2 Numerical Method

The subcell limiter algorithm of Hennemann et al. [3] is described here in brief for the one-dimensional Euler equations

$$\frac{\partial \boldsymbol{\mathcal{U}}}{\partial t} + \frac{\partial \boldsymbol{\mathcal{F}}}{\partial x} = 0, \quad \boldsymbol{\mathcal{U}} = [\rho, \rho u, \rho E]^T, \quad \boldsymbol{\mathcal{F}} = [\rho u, \rho u^2 + p, (\rho E + p)u]^T, \tag{8.1}$$

where ρ, u and p denote the density, velocity and pressure respectively, and $E := p/((\gamma - 1)\rho) + u^2/2$ is the total specific energy with γ being the specific gas constant for the perfect gas model considered here. Consider the restriction of Eq. (8.1) to a representative element $\Omega_e := [a, b]$, which is being mapped to the reference element $\Omega_{\text{ref}} := [0, 1]$ using the linear mapping $\xi \overset{\mathcal{M}}{\longmapsto} x$:

$$\mathcal{M}^{-1}(x) = \xi = \frac{x - a}{b - a} \quad (x \in \Omega_e), \quad J\frac{\partial \mathbf{U}}{\partial t} + \frac{\partial \mathbf{F}}{\partial \xi} = 0 \quad (\xi \in \Omega_{\text{ref}}),$$

where $\mathbf{U}(\xi, t) := \boldsymbol{\mathcal{U}}(x, t)$, $\mathbf{F}(\mathbf{U}) := \boldsymbol{\mathcal{F}}(\mathbf{U})$ are the conservative state and flux in reference space, and $J := \mathrm{d}\mathcal{M}/\mathrm{d}\xi = (b - a)$ is the Jacobian of the mapping. $\mathbf{U}$ and $\mathbf{F}$ are approximated to lie in a polynomial space of N-th degree polynomials

$$\mathbf{U}(\xi,t) \approx \sum_{i=0}^{N} \mathbf{U}_i(t)\ell_i(\xi), \quad \mathbf{F}(\xi,t) \approx \sum_{i=0}^{N} \mathbf{F}(\mathbf{U}_i(t))\ell_i(\xi),$$

where $\{\ell_i(\xi)\}_{i=0}^N$ are the basis functions of the polynomial space, and $\{\mathbf{U}_i\}_{i=0}^N$ are the nodal values of the solution. These basis functions are the Lagrange functions constructed using the $(N+1)$ Gauss-Lobatto (GL) points $\{\xi_i\}_{i=0}^N$ (termed "nodes"). The error in this approximation is minimised using the Galerkin approach in a strong sense for the test functions $\{\ell_j\}_{j=0}^N$:

$$\int_0^1 \ell_j J \frac{\partial \mathbf{U}}{\partial t}\,\mathrm{d}\xi + \int_0^1 \ell_j \frac{\partial \mathbf{F}}{\partial \xi}\,\mathrm{d}\xi + \left[\ell_j(\mathbf{F}^* - \mathbf{F})\right]_{\xi=0}^{\xi=1} = 0 \quad (j = 0, 1, \ldots, N), \tag{8.2}$$

where $\mathbf{F}^*$ is the numerical flux function introduced at the element boundaries to ensure conservation. Unless otherwise mentioned, the range of the indices i and j is 0 to N in all the following equations. Inexact integration of the integrals in Eq. (8.2) is performed using **colocated quadrature**, i.e., using GL quadrature with $(N+1)$ points. This yields the standard (abbr. "std") discretisation of the governing equation

$$-J\dot{\mathbf{U}}_i = \sum_{j=0}^{N}\left[D_{ij}\mathbf{F}_j\right] - \frac{\delta_{0i}}{w_0}\left(\mathbf{F}_0^* - \mathbf{F}_0\right) + \frac{\delta_{Ni}}{w_N}\left(\mathbf{F}_N^* - \mathbf{F}_N\right) := \mathsf{R}_{\mathrm{std}}(\mathbf{F})_i \tag{8.3}$$

where $D_{ij} = \ell_j'(\xi_i)$, and $\mathsf{R}_{\mathrm{std}}$ is used to denote the residual due to standard DG discretisation. The use of colocated quadrature also allows the volumetric term of Eq. (8.3) to be written as a flux difference:

$$\sum_{j=0}^{N} D_{ij}\mathbf{F}_j = \frac{\widehat{\mathbf{F}}_{i+1} - \widehat{\mathbf{F}}_i}{w_i}, \tag{8.4}$$

where $\{w_i\}_{i=0}^N$ are the GL quadrature weights, and $\{\widehat{\mathbf{F}}_i\}_{i=0}^{N+1}$ are the corresponding consistent volumetric fluxes defined on the **flux points** $\{\widehat{\xi}_i\}_{i=0}^{N+1}$. The flux points are related to the GL points and the GL weights: $\widehat{\xi}_0 = \xi_0, \widehat{\xi}_{N+1} = \xi_N$ and $\widehat{\xi}_{i+1} - \widehat{\xi}_i = w_i$. This results in a decomposition of every element into subcells of widths $\{w_i\}_{i=0}^N$ in the reference space as shown in Fig. 8.1. With the help of Fig. 8.1, one can see that Eq. (8.4) provides an interpretation of the standard DG discretisation as a subcell finite volume (FV) discretisation with a flux reconstruction.

For anti-aliasing and entropy/kinetic energy stability in smooth regions, a **split-form** (abbr. "sf") DG discretisation is proposed by setting the volumetric fluxes $\{\widehat{\mathbf{F}}_i\}_{i=0}^{N+1}$ as

$$\begin{aligned} &\widehat{\mathbf{F}}_0^{\mathrm{HO}} \approx \mathbf{F}_0, \quad \widehat{\mathbf{F}}_{N+1}^{\mathrm{HO}} \approx \mathbf{F}_N, \\ &\widehat{\mathbf{F}}_i^{\mathrm{HO}} \approx \sum_{j=i}^{N}\sum_{k=0}^{i-1} 2w_k D_{kj}\mathbf{F}^{\#}(\mathbf{U}_k, \mathbf{U}_j) \quad (i = 1, 2, \ldots, N), \end{aligned} \tag{8.5}$$

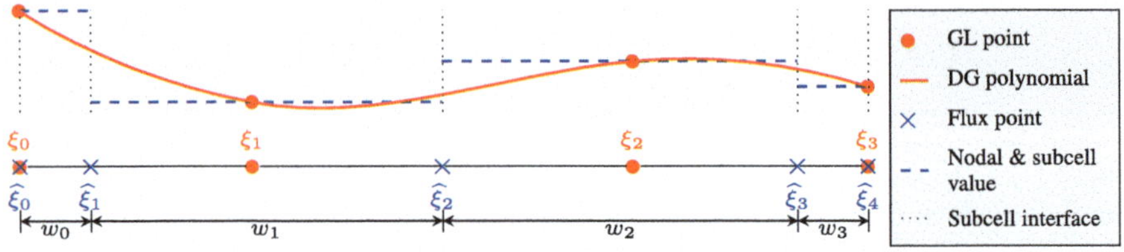

Fig. 8.1 Equivalent subcell FV decomposition for a DGSEM element with $N = 3$

where "HO" abbreviates "high order" and $\mathbf{F}^{\#}(\bullet, \bullet)$ is a **volumetric two-point flux function**. Substituting Eq. (8.5) and (8.4) in Eq. (8.3) gives the split-from discretisation.

$$-J\dot{\mathbf{U}}_i = 2\sum_{j=0}^{N} D_{ij}\mathbf{F}^{\#}(\mathbf{U}_i, \mathbf{U}_j) - \frac{\delta_{0i}}{w_0}\left(\mathbf{F}_0^* - \mathbf{F}_0\right) + \frac{\delta_{Ni}}{w_N}\left(\mathbf{F}_N^* - \mathbf{F}_N\right) := \mathsf{R}_{\mathrm{sf}}^{\mathrm{HO}}(\mathbf{F})_i \tag{8.6}$$

Limiting is achieved by blending this split-form discretisation with a low order discretisation. Since the introduction of flux points and volumetric fluxes provides a natural subcell decomposition of the finite element (see Fig. 8.1), the low order (abbr. "LO") scheme is constructed by defining the volumetric fluxes using the left and right subcell values.

$$\widehat{\mathbf{F}}_0^{\mathrm{LO}} \approx \mathbf{F}_0, \quad \widehat{\mathbf{F}}_{N+1}^{\mathrm{LO}} \approx \mathbf{F}_N, \quad \widehat{\mathbf{F}}_i^{\mathrm{LO}} \approx \mathbf{F}^*(\mathbf{U}_{i-1}, \mathbf{U}_i) \quad (i = 1, 2, \ldots, N). \tag{8.7}$$

Substituting Eqs. (8.7) and (8.4) in Eq. (8.3) gives the low order subcell (abbr. "sc") discretisation.

$$-J\dot{\mathbf{U}}_i = \frac{\widehat{\mathbf{F}}_{i+1}^{\mathrm{LO}} - \widehat{\mathbf{F}}_i^{\mathrm{LO}}}{w_i} - \frac{\delta_{0i}}{w_0}\left(\mathbf{F}_0^* - \mathbf{F}_0\right) + \frac{\delta_{Ni}}{w_N}\left(\mathbf{F}_N^* - \mathbf{F}_N\right) := \mathsf{R}_{\mathrm{sc}}^{\mathrm{LO}}(\mathbf{F})_i \tag{8.8}$$

The final semi-discrete form is obtained by blending the high order split-form (Eq. 8.6) and the low order subcell form (Eq. 8.8) discretisations using a factor α.

$$J\dot{\mathbf{U}}_i + (1-\alpha)\mathsf{R}_{\mathrm{sf}}^{\mathrm{HO}}(\mathbf{F})_i + \alpha\mathsf{R}_{\mathrm{sc}}^{\mathrm{LO}}(\mathbf{F})_i = 0 \tag{8.9}$$

An indicator that estimates the relative energy in high order modes of the quantity $\varepsilon = p\rho$ is used to calculate α, and its value is capped at a parameter $\alpha_{\mathrm{max}} = 0.5$. A high value of α gives a higher low order contribution and hence stabilises the solution. Equation (8.9) is integrated in time using an explicit low-storage 5-stage RK4 time integrator [4] with the following time step restriction [5]

$$\Delta t_e = \mathrm{CFL} \times \frac{1}{N^{1.5}} \min_{\text{Element nodes}} \left[\frac{1}{\left\|\underline{\mathbf{J}}^{-1}\mathbf{u}\right\|_\infty + a/h_e}\right], \quad \Delta t = \min_{\text{Elements}} \Delta t_e, \tag{8.10}$$

where $\|\bullet\|_\infty$ denotes the infinite vector norm, $\underline{\mathbf{J}}$ is the Jacobian matrix for mapping a three-dimensional curved hexahedral element to the reference element, $\mathbf{u}$ is the velocity vector, and h_e is the minimum vertex distance for an element. Chandrashekhar volume flux and numerical flux [3] are used for $\mathbf{F}^{\#}$ and $\mathbf{F}^{*}$ respectively. Boundary conditions are implemented in a "weak-Riemann" sense by setting $\mathbf{F}^{*}$ at boundaries [6]. The three-dimensional version of the limiter is implemented in C++ using the `deal.II` finite element library [7].

8.3 Results and Discussion

Three metrics are considered to quantify the performance of a simulation: (i) Memory requirement (ii) Cost, and (iii) Accuracy. For a DG method, the Degree of Freedom (DoF) count determines the Memory requirement. A finite element using N-th order polynomial basis has $(N+1)^d$ DoFs, where d is the space dimension of the element. Consequently, on a given mesh, higher polynomial degree would have a higher total DoF count. Therefore, for a fair comparison, simulations using different polynomial degree are performed on different meshes, designed such that the DoF count remains (nominally) constant. As regards the Cost, CPU time/time step of the simulation can be considered a reasonable estimate.[1] To estimate the Accuracy, a reference solution—preferably analytical—is generally required. Since the shock interaction problems considered in this work (Sects. 8.3.2 and 8.3.3) do not have an analytical solution, a one-dimensional shock tube problem is considered in Sect. 8.3.1 to perform an Accuracy vs Cost analysis. This would provide a reliable basis to comment on the effect of N on the performance of DG simulations (in terms of accuracy and cost), albeit, for that specific case. Moreover, the Cost/DoF estimates noted in Sect. 8.3.1 can be used even for the subsequent shock interaction cases. Air is the fluid used for all simulations.

8.3.1 *One-dimensional Riemann Problem*

The modified Sod shock tube case [8] is considered here for Accuracy vs Cost comparison. The initial condition is $(\rho, u, p)_L = (1, 0.75, 1)$ and $(\rho, u, p)_R = (0.125, 0, 0.1)$ with the initial discontinuity at $x = 0.3$. The simulations are performed for $N = 1, 2, 3, 5$ on a three-dimensional domain $[0, 1] \times [0, 0.5]^2$ with $\mathrm{lcm}\{(N+1) \mid N = 1, 2, 3, 5\} = 12$ DoFs in the y and z directions to ensure integral number of elements in these directions. Three values of DoFs in the x direction are considered: 200, 400 and 800. All simulations use CFL $= 0.5$. Exact boundary conditions are used on left and right, while symmetry boundary conditions are used on the remaining sides.

---

[1] It must be noted that Eq. (8.10) generally gives a lower time step value for higher N at a given DoF resolution.

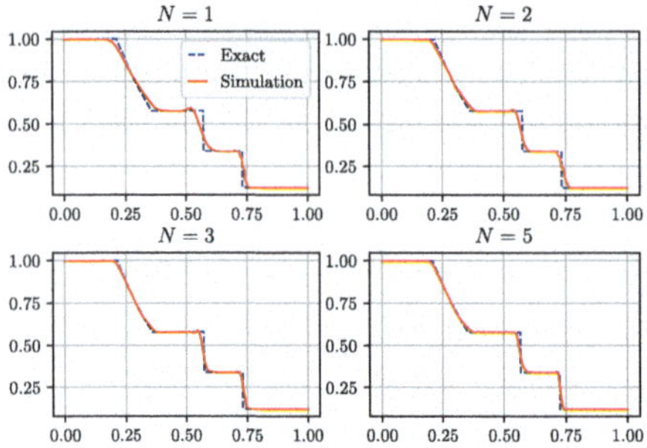

(a) Visual comparison for 200 DoFs.

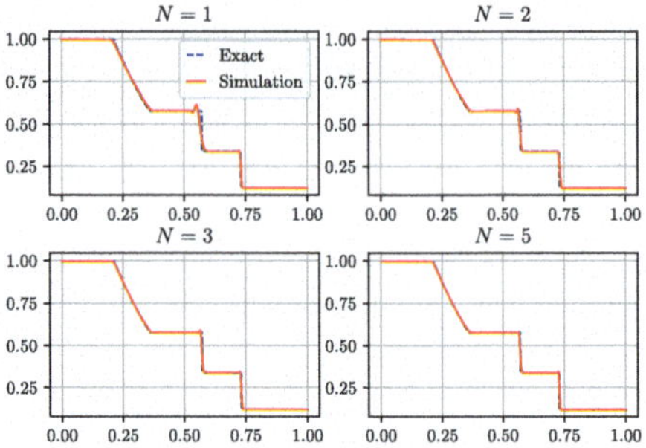

(b) Visual comparison for 800 DoFs.

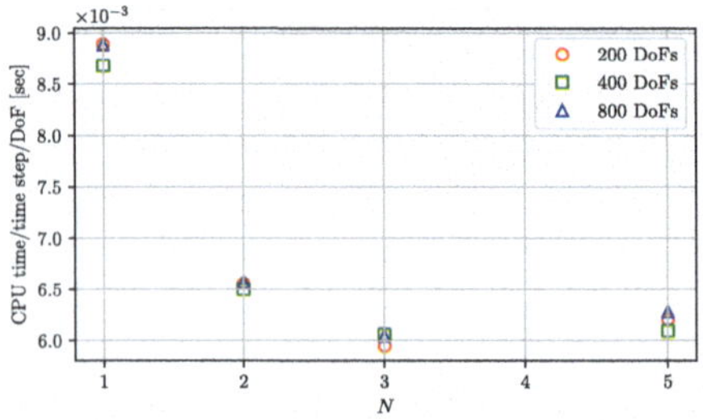

(c) Cost/DoF vs N.

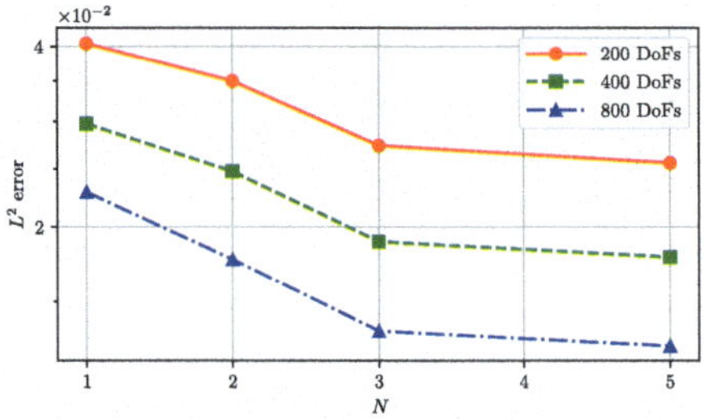

(d) Error vs N.

Fig. 8.2 Results for the modified Sod shock tube case. Density is the variable used for comparison and analysis

Figure 8.2 shows the results obtained on 6400 sampling points along the centreline of the three-dimensional domain. CPU time is obtained using a `deal.II`'s built-in function, and L^2 error is computed using the numerical and analytical density values at the sampling points. Figure 8.2a, b show the density variation obtained at the lowest (200) and highest (800) DoF resolution. Figure 8.2a shows that the resolutions of shock (at $x \approx 0.75$) and the contact wave (at $x \approx 0.6$) improve with N until $N = 3$, and only have a small improvement with $N = 5$. At 800 DoF resolution, Fig. 8.2b shows that while the shock resolution is the same for all values of N, the contact wave resolution is better with $N = 2, 3, 5$. Figure 8.2c shows that the Cost/DoF has similar value for $N = 2, 3, 5$ with a minimum at $N = 3$, while $N = 1$ has approximately a 1.5 times higher value. In general, DG algorithms are not computationally favourable for second order simulations ($N = 1$) in terms of Memory requirement and Cost. It must be mentioned here that these Cost estimates are very specific to the current implementation of the algorithm. Figure 8.2d shows that $N = 3, 5$ have about 1.5 times lower error compared to $N = 1$. On the whole, it can be concluded that for this particular case, increasing N (keeping the DoFs fixed) has a positive outcome until $N = 3$, and the drop in error from $N = 3$ to $N = 5$ is only marginal, with a slight increase in Cost.

8.3.2 One-dimensional Shock Interaction Problem

Here, the well-known Shu-Osher test case [9] is considered, which consists an interaction of a right moving Mach 3 shock with a smooth density variation. An initial discontinuity at $x = -4$ is imposed in a domain $[-5, 5] \times [0, 1]^2$, with $(\rho, u, p) = (3.857\,143, 2.629\,369, 10.333\,33)$ and $(\rho, u, p) = (1 + 0.2 \sin 5x, 0, 1)$ to the left and right of the discontinuity respectively. The end time is $t = 1.8$ s. Like in Sect. 8.3.1, $N = 1, 2, 3, 5$ are considered for (nominal) DoF resolutions 200, 400 and 800 in the x direction. The y and z directions are meshed with a single cell and hence have different DoFs for different values of N. A third order ENO solution on 1600 points by Shu and Osher [9] is taken as the reference for comparison.

Figure 8.3 shows the results obtained on 1600 sampling points, along with insets which show a focused view of the variation behind the shock. First, it is noted that at all DoF resolutions, solutions using $N > 1$ perform marginally better than $N = 1$ in capturing the high frequency variation behind the shock (see insets). There is not much difference among the solutions using $N = 2, 3, 5$ in this particular post-shock region. Increasing N has a positive effect on the thickness of shock (at $x \approx 2.5$) at lower DoF resolutions (Fig. 8.3a, b). At the highest (800) DoF resolution, while the shock thickness itself looks independent of N, the transition from high frequency waves just behind the shock to the shock jump is increasingly well predicted with higher values of N (see the variation just left to the shock in insets of Fig. 8.3c). The low frequency waves behind (to the left of) high frequency waves develop into weak shocks, and all simulations produce oscillations at these locations since the limiter is inactive here. The magnitude of these oscillations is observed to marginally decrease with N at all DoF resolutions. Together with the Cost estimates noted in Fig. 8.2c, it can be concluded that using a higher polynomial degree indeed has a benefit in this case.

8.3.3 Two-dimensional Shock Interaction Problem

The double Mach reflection test case [10] is considered here, which involves a Mach 10 shock hitting a ramp angled at 30° relative to the direction of shock motion. Both the incident and the reflected shocks undergo a Mach reflection producing a complex wave structure. Following Kemm [11], the problem is setup in the domain $[0, 4] \times [0, 2]$ with the incident shock initially inclined 60° to the wall, which is one part ($x \in [1/6, 4]$) of the bottom boundary. The other part of the bottom boundary ($x \in [0, 1/6]$) along with the left boundary is given post-shock inflow conditions. The top side of the domain is given an exact boundary condition to capture the motion of the shock, and the right boundary is given the pre-shock condition: $p = 1$, $\rho = \gamma$, and $u = 0 = v$. The meshes used in this case are designed to have $\Delta x = \Delta y \approx (N + 1)/120$ for $N = 1, 2, 3, 5$. Since it is necessary to have integral number

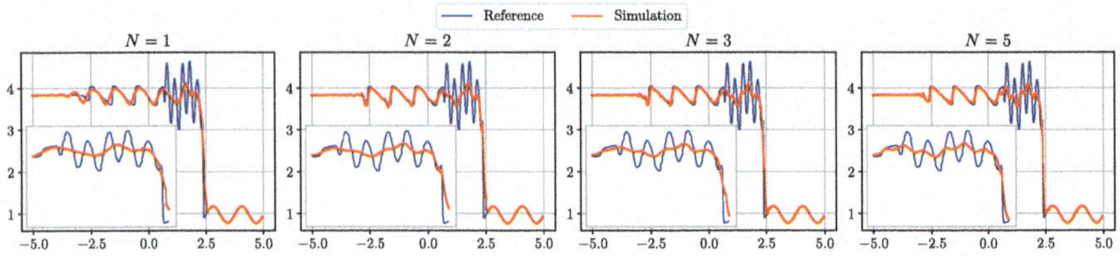

(a) 200 DoFs.

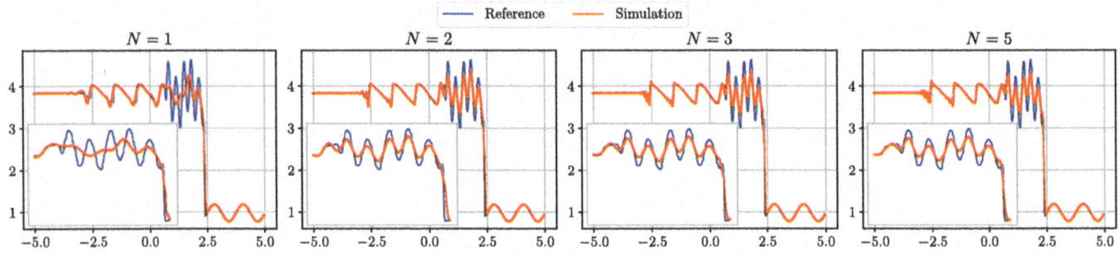

(b) 400 DoFs.

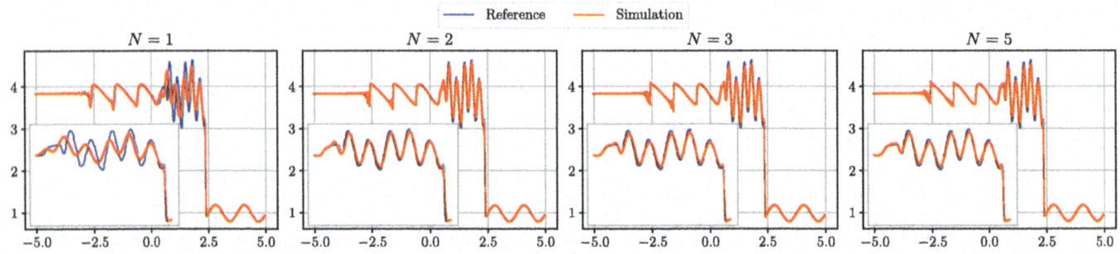

(c) 800 DoFs.

Fig. 8.3 Shu-Osher test results. Reference solution is a third order ENO solution on 1600 points by Shu and Osher [9]. Insets in all figures show a close-up view in the range $x \in [0.3, 2.5]$

Table 8.1 DoF count for different meshes used in the double Mach reflection case

N	1	2	3	5
DoFs	115 200	115 920	115 200	116 640

of cells meshing both parts of the bottom boundary, the DoF count is not exactly the same in all these cases, as listed in Table 8.1.

Figure 8.4 shows 32 density contours, plotted over a surface plot of vertical momentum (chosen since it best depicts the secondary slip line (s') [10]). To set a context for discussion, Fig. 8.4f shows a focused view of the main interaction region along with a marking of the different waves. Figure 8.4a–d use the range limits corresponding to $N = 1$ case for the contour and surface plots. A brown patch in Fig. 8.4b, and spurious density contours in the pre-shock region of Fig. 8.4b, d are both due to a mismatch with the ranges so chosen.

Figure 8.4 shows that the incident shock (i) and the secondary Mach stem (m') display a staircase-like pattern for $N = 2, 3$, while they are more smoothly resolved with $N = 1, 5$. Thicknesses of the primary Mach stem (m) and (i) are similar in

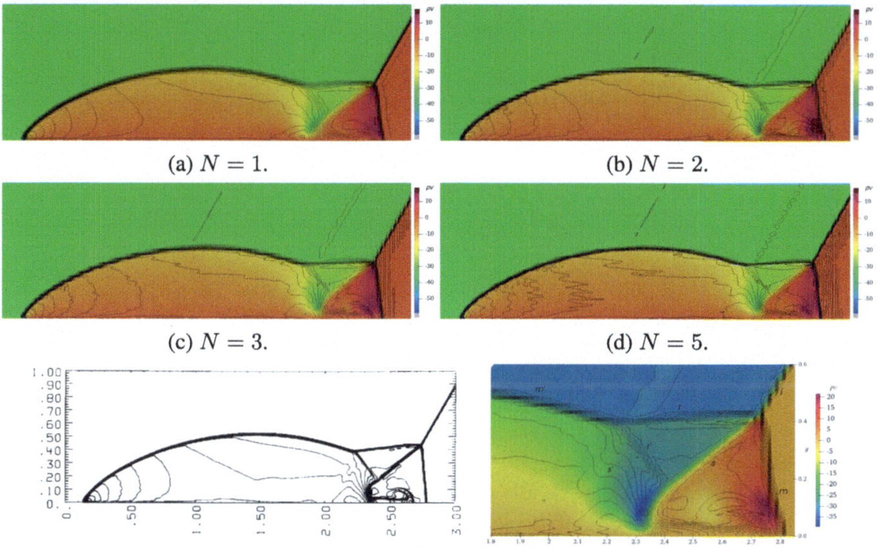

(a) $N = 1$.

(b) $N = 2$.

(c) $N = 3$.

(d) $N = 5$.

(e) Reference solution using PPMDE and $\Delta x = \Delta y = 1/120$ [10].

(f) Close-up view of (b) along with a demarcation of the wave structure.

Fig. 8.4 Results of the double Mach reflection test case. Figures **a** to **d** show vertical momentum surface plot in the range $[-59, 16]$. Values below and above this range are displayed in grey and brown respectively. 32 density contours in the range $[1.4, 21]$ are drawn over this surface plot. Notation used in Fig. (**f**): (i)–incident shock, (m/m')–primary/secondary Mach stem, (r/r')–primary/secondary reflected shock, (s/s')–primary/secondary slip line, (T/T')–primary/secondary triple point

all cases, while the resolutions of (m') and primary reflected shock (r) are better when $N > 1$. This is also distinctly visible in Fig. 8.5a which plots density variation along a line at $x = 2.5$. Similarly, resolution of the primary slip line (s) is better with $N = 2, 3, 5$ compared to $N = 1$, as confirmed again in Fig. 8.5a, and also in Fig. 8.5b which shows the density variation along $y = 0.3$ line. As regards the bending of (s) near the wall, while $N = 3, 5$ perform better in capturing this phenomenon, $N = 2$ shows a slightly different result as it pushes the primary Mach stem (m) forward. Furthermore, $N = 1$ fares poorly compared to $N = 2, 3, 5$ in this regard. The secondary reflected shock (r') is quite smeared in $N = 1$ case, while it is better resolved at higher N values. This can be very clearly inferred from Fig. 8.5b. None of the cases is able to sharply capture the secondary slip line (s'), although $N = 2, 3, 5$ perform marginally better than $N = 1$ in this aspect, as can be seen in Fig. 8.5c (showing vertical momentum variation along $y = 0.2$ line), and Fig. 8.5b. To conclude, one can say that $N = 2, 3, 5$ perform better than $N = 1$ in this case, although there is no significant improvement in results beyond $N = 2$. A minor benefit in using $N = 3$ over $N = 2$ is its marginally lower Cost, as shown in Fig. 8.2c.

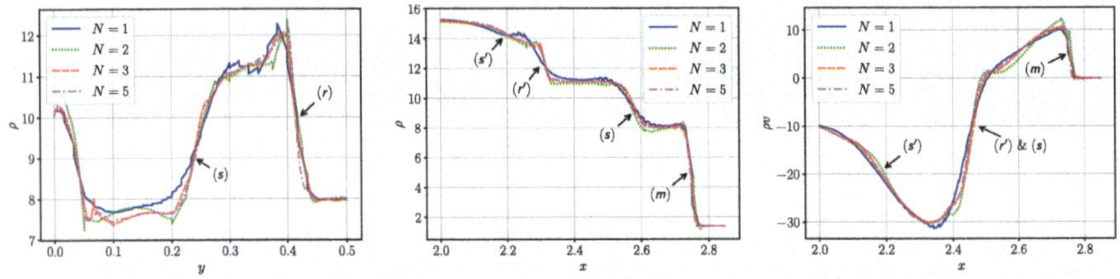

(a) $x = 2.5$. (b) $y = 0.3$. (c) $y = 0.2$.

Fig. 8.5 Variation of quantities on certain sampling lines for the double Mach reflection test case

8.4 Conclusions

This work was aimed at investigating the effect of polynomial degree N on DG simulation of Euler equations. A subcell limiter, which blends a low order FV scheme with the high order DG scheme is used for analysis. To quantify the performance of a simulation, three metrics have been used: (i) DoF count (Memory requirement), (ii) CPU time/time step (Cost), and (iii) Accuracy. In all cases, simulations using different values of N have been performed by fixing the DoF count. For the modified Sod shock tube problem (which has an analytical solution), a Cost vs Accuracy analysis revealed that while simulations using higher values of N perform better, the improvement beyond $N = 3$ is only marginal. It was also noted here that Cost/DoF show a minimum $N = 3$, with almost similar values for $N = 2, 5$. In the Shu-Osher test, it was again noted that higher polynomial degrees perform better, with no such apparent performance plateau. Interestingly, in these two one-dimensional cases, it was observed that the accuracy gain with increasing N is more pronounced at lower DoF resolutions. On the other hand, it was observed in the double Mach reflection case that although simulations using $N = 2, 3, 5$ perform better than $N = 1$, there is no practical benefit in going beyond $N = 2$.

It can be concluded that high order simulations using $N > 1$ perform better than $N = 1$ in all the cases considered. However, in flows dominated by strong discontinuities, the accuracy gain plateaus at $N = 2$. On the other hand, in problems dominated by smooth variation and/or weak discontinuities, this plateau is pushed to a higher value of N. These conclusions encourage the development of hp-adaptive techniques for compressible flows that use a quadratic basis ($N = 2$) at (strong) discontinuities and a high order basis ($N \approx 5$) in smooth regions. A cubic basis ($N = 3$) may be used at weak discontinuities. The corresponding h-adaptation can be appropriately done to keep the total DoF count constant. Conducting complementary studies and assessing the validity of these conclusions on an extended set of two-dimensional test cases would be part of a future work. Performing a similar study for laminar and turbulent compressible flows containing discontinuities can also be an interesting extension of this work.

References

1. J. Hesthaven, T. Warburton, Springer Sci. Bus. Media (2007). https://doi.org/10.1007/978-0-387-72067-8
2. B. Cockburn, C.W. Shu, J. Sci. Comput. **16**(3), 173–261 (2001). https://doi.org/10.1023/A:1012873910884
3. S. Hennemann et al., J. Comput. Phys. **426**, 109935 (2021). https://doi.org/10.1016/j.jcp.2020.109935
4. C.A. Kennedy, M.H. Carpenter, R. Michael Lewis, Appl. Numer. Math. **35**(3), 177–219 (2000). https://doi.org/10.1016/S0168-9274(99)00141-5
5. M. Kronbichler, Last accessed 2022/04/19. URL: https://www.dealii.org/current/doxygen/deal.II/step67.html
6. G. Mengaldo et al, in *AIAA AVIATION 2014 -7th AIAA Theoretical Fluid Mechanics Conference*. (2014). https://doi.org/10.2514/6.2014-2923.
7. D. Arndt et al., J. Numer. Math. **29**(3), 171–186 (2021). https://doi.org/10.1515/jnma-2021-0081
8. E.F. Toro, Third edition. Springer-Verlag Berlin Heidelberg, (2009). https://doi.org/10.1007/b79761
9. C.-W. Shu, S. Osher, J. Comput. Phys. **83**(1), 32–78 (1989). https://doi.org/10.1016/0021-9991(89)90222-2
10. P. Woodward, P. Colella, J. Comput. Phys. **54**(1), 115–173 (1984). https://doi.org/10.1016/0021-9991(84)90142-6
11. F. Kemm, Computers & Fluids **132**, 72–75 (2016). https://doi.org/10.1016/j.compfluid.2016.04.008

Chapter 9
Numerical Studies on the Aerodynamics of a High-Speed Spinning Projectile

Rajkamal Sah and Gopalan Jagadeesh

Abstract The projectile during travel inside gun barrel, after firing, has varying flow physics over it from subsonic to supersonic regime. The quantitative investigation of boundary layer over the spinning cone was done numerically by changing different parameters using Ansys fluent code. The parameters considered for study was Mach number, Reynolds number as well as Strouhal number. Boundary layer thickness as well as pressure increases with increasing Strouhal number. The decrease in Reynolds number increases the boundary layer thickness up to 1 mm. Effect of spin at high Mach number is within 0.05 mm whereas the density and velocity were seen to be varied up to 0.5 mm thickness.

9.1 Introduction

When gun is fired, the projectile travels via gun barrel to muzzle exit. The launch of projectiles can be divided into two distinct phases: the confined volume acceleration and the free flight that are also known as set-back and set-forward phases, respectively. The transition from confined volume to free flight is known as the muzzle exit [1]. The flow as well as pressure around the projectile will develop and change from subsonic regime to supersonic regime based on desired muzzle velocity [2] (Fig. 9.1). Many experiments and studies were done in past which investigated the external aerodynamic study and ways to get stability on the spinning projectile [3–7]. Interior ballistic process was ignored or simply assumed in most investigations [2]. The skin friction, pressure drag, and base drag are the source of drag on the projectile body [8].

The effect of spin on boundary layer as well as shock profile is not quantified as they play significant role in internal aerodynamics as bullet travels via gun barrel. The study on spinning cone was done in past [3–7]. But it does not cover the full regime from subsonic to supersonic. The intention of present paper is to present the numerical simulation quantitative result of boundary layer studies of spinning cone from subsonic to supersonic regime as spinning affects the boundary layer profile.

R. Sah · G. Jagadeesh (✉)
Department of Aerospace Engineering, Indian Institute of Science, Bengaluru, India
e-mail: jaggie@iisc.ac.in

© The Author(s), under exclusive license to Springer Nature Singapore Pte Ltd. 2024
G. Rajesh et al. (eds.), *Advances in Shock Interactions*, Springer Proceedings in Physics 408, https://doi.org/10.1007/978-981-97-6099-2_9

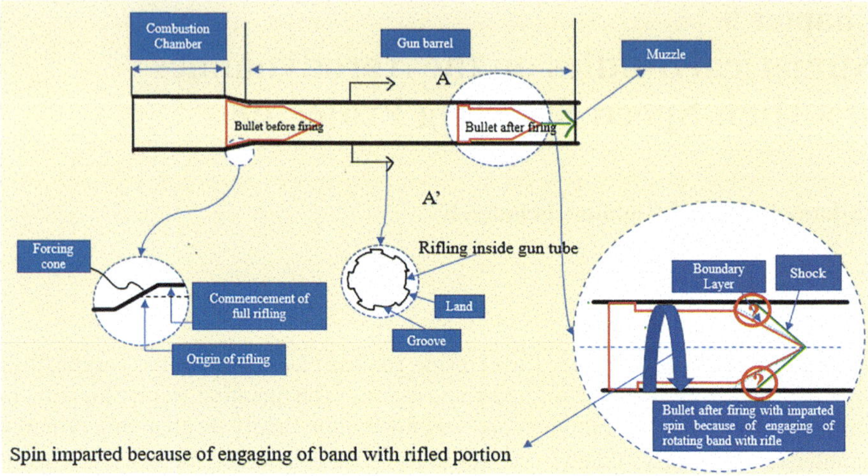

Fig. 9.1 Schematics of inside gun barrel of how the bullet travels inside it after firing **a**Showing bullet inside gun barrel before firing and after firing respectively **b** part of barrel where rotating band of bullet resides and rifling starts **c** inner portion of gun barrel having rifling in the form of lands and groove **d** zoomed view of projectile travelling inside gun barrel and shock as well as boundary layer formation

9.2 Methodology

9.2.1 Geometry Definition and Flow Parameters

Cone body represents the simplest projectile because in the case of rotating cones one can investigate changes in the boundary-layer characteristics caused by the rotational motion by using a small number of geometrical and mechanical parameters. The cone diameter considered was 50 mm and half angle was 15° (Fig. 9.2). The flow over cone can be considered axisymmetric at zero-degree angle of attack. Axisymmetric form of Navier–stokes flow equations was used in cylindrical coordinate. The frame of reference is shown in Fig. 9.2, and the computational domain for the simulation is depicted in Fig. 9.3.

The flow parameters considered after dimensional analysis of governing equations used was Reynolds number (Re), Mach number (M) as well as Strouhal number (Sr).

9.2.2 Grid Generation of Flow Domain

Two-dimensional axisymmetric domain was made in Ansys design modeler. Structured meshing was done in ICEM CFD for dividing the flow domain into small domain. Three-dimensional structured meshing was also done for matching the result

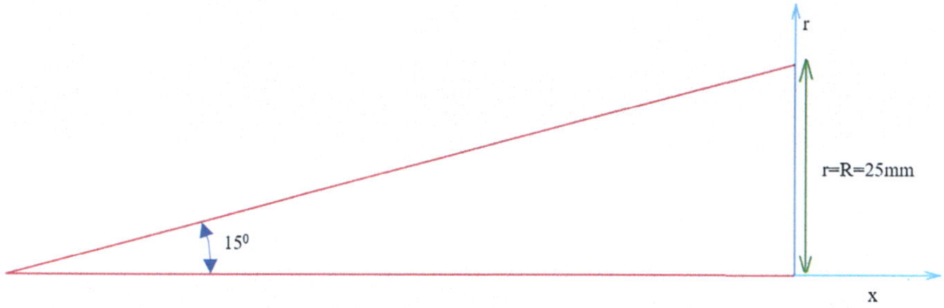

Fig. 9.2 Geometry considered for flow investigation over cone

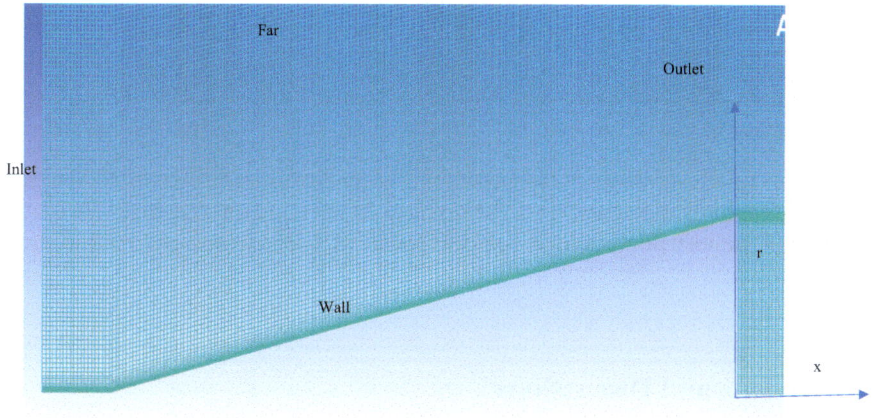

Fig. 9.3 Axisymmetric flow mesh domain for conical body

using this with axisymmetric result (Fig. 9.4). Initially 0.4 million elements were considered for study. Grid independence was performed using five x meshes staring from 0.4 million to 6.4 million in steps of 2 times from previous. After optimum size of mesh was found (2.6 million) we used density based Laminar model for steady as well as unsteady flow analysis. Ideal gas law was used for working fluid.

9.2.3 Numerical Method Outline

The parameters considered for study include was the Mach number, Reynolds number and Strouhal number. The data for boundary layer thickness was compared with result of reference [9] and it is matching well. The projectile non dimensional spin rate (ω d/2v) range considered was 0.05 to 0.5. Laminar viscous model was used based on the previous study. A standard three-coefficient Sutherland viscosity model was used. The freestream density was set to match experimental conditions. The range

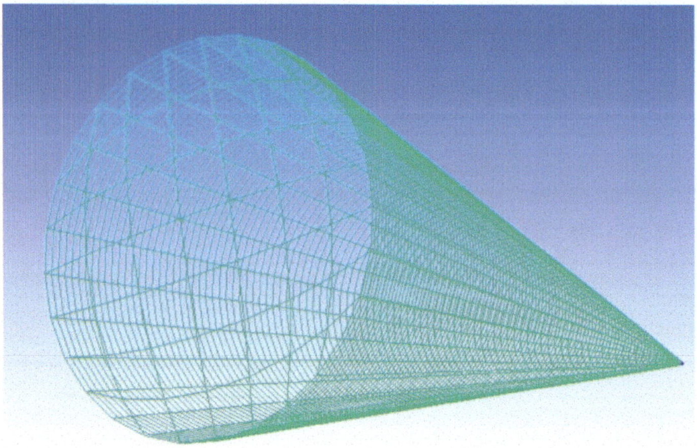

Fig. 9.4 Three-dimensional structured grid for conical body generated in ICEM CFD

of Reynolds number variation considered for the study was 26 million to 2600 based on model length. The variation of Mach 0.3 to Mach 2 was investigated. A 64-bit commercial, finite-volume RANS solver (Fluent 2021 R2) was used to produce all the results.

9.3 Results and Discussions

9.3.1 Strouhal Number Change

Mach 2 flow was considered and Strouhal number ranging from zero to 0.5 was taken into consideration. The location in flow domain considered for generating the result was along perpendicular line (length = 3 mm) from cone surface at 40 mm from cone base towards cone tip and. It was found that spin distorted the velocity boundary layer thickness, density as well as pressure profile as shown in Figs. 9.5, 9.6, 9.7 and 9.8. Velocity first decreases from cone surface velocity to minimum and then it increases to match the free stream velocity (Fig. 9.5) and this change in properties was more pronounced with higher Strouhal number (Fig. 9.6). Figure 9.8 shows the pressure profile result. Near wall pressure first increase up to some height then decrease and again decrease uniformly after slight increasing its value.

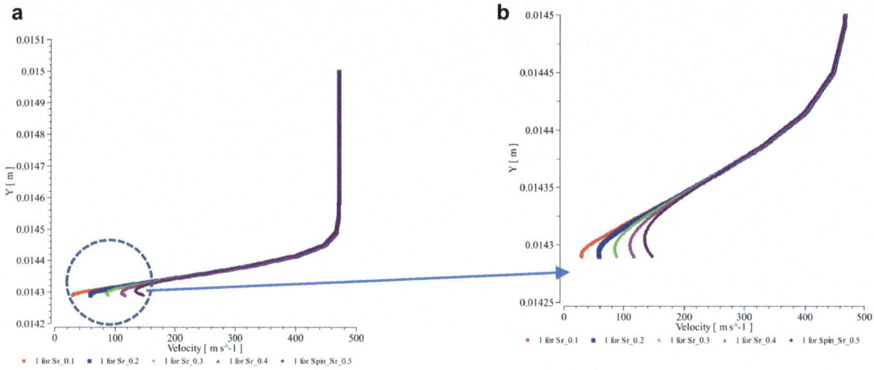

Fig. 9.5 **a** Velocity plot comparison at different Strouhal number from value zero to 0.5. **b** Velocity plot zoomed portion near the wall

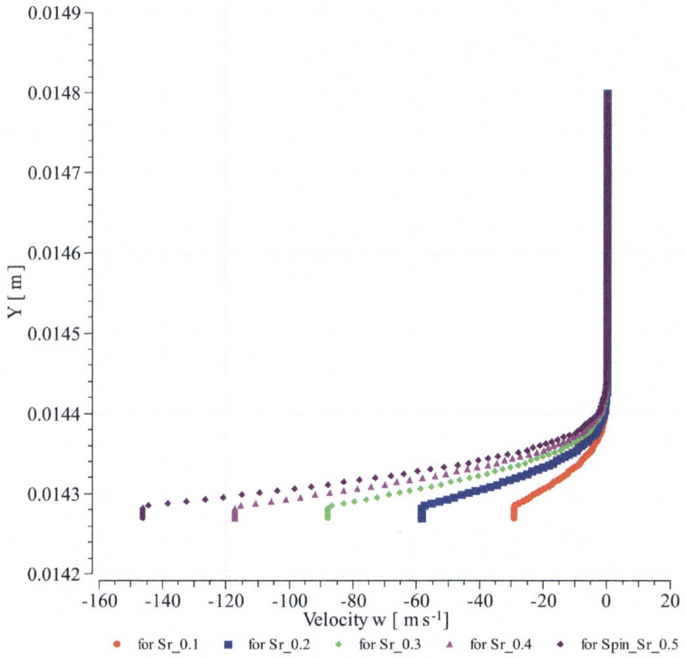

Fig. 9.6 Tangential velocity plot for Mach 2 at different Strouhal number

9.3.2 Reynolds Number Change

During travel of projectile via gun barrel its Reynolds number also changes. The intention of the current investigation is to quantify the effect of Reynolds number change. Figure 9.9 shows the velocity contour comparison with four different

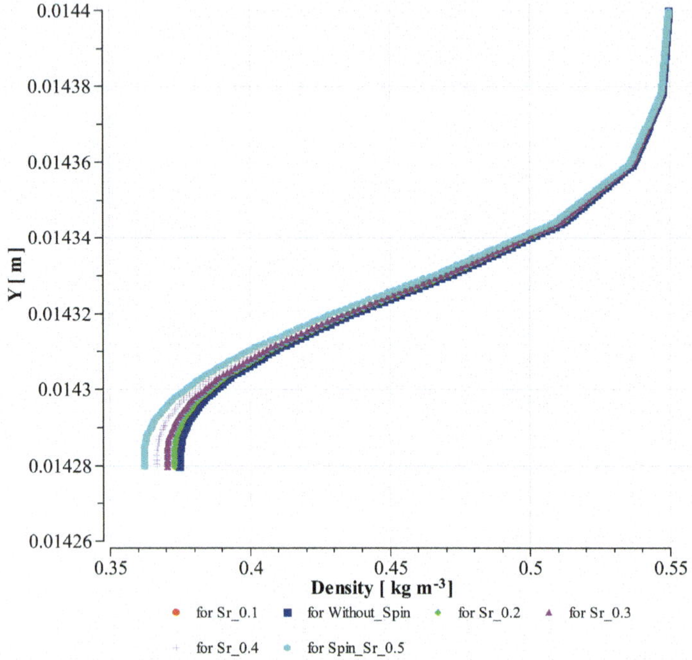

Fig. 9.7 Density plot for Mach 2 flow at different Strouhal number

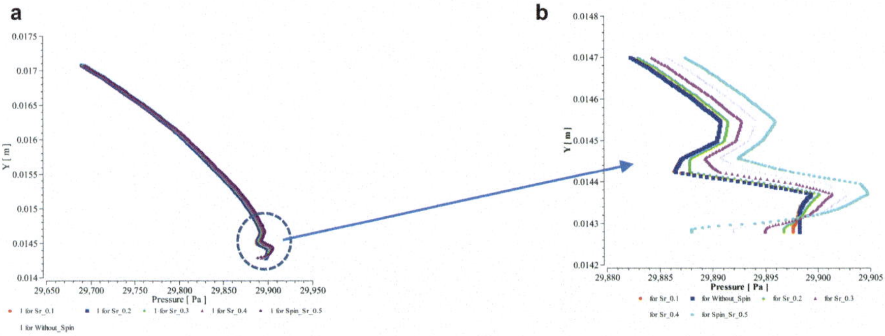

Fig. 9.8 **a** Pressure plot at Mach 2 with different Strouhal number comparison. **b** Pressure plot comparison zoomed view for clearly visualization of near wall region

Reynolds number i.e., 2.6e6 to 2.6e3 in steps of one order reduction when projectile does not spin. It was found that the velocity boundary layer near cone surface was getting thicker as Reynolds number got decreased (Fig. 9.9). The effect of spin at lower Reynolds number can also be seen by comparison of Figs. 9.10 and 9.12.

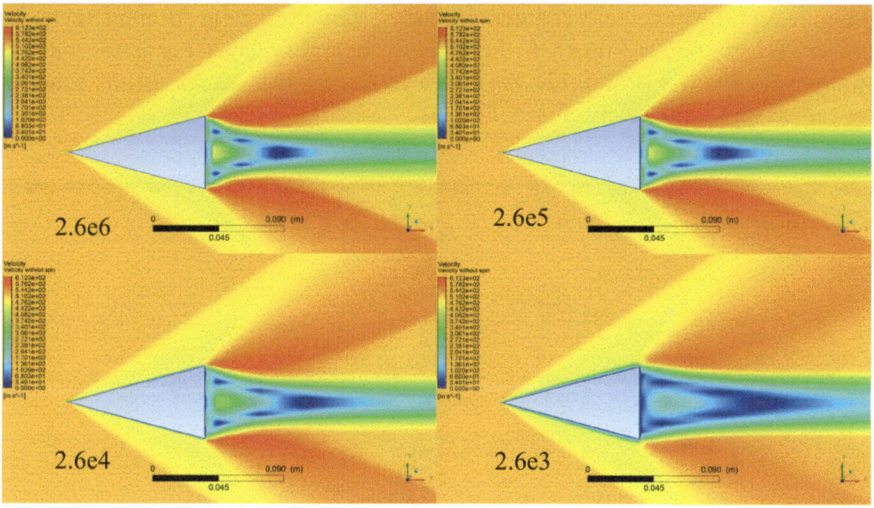

Fig. 9.9 Velocity contour for Reynolds number 1e6 to 1e3 without spin

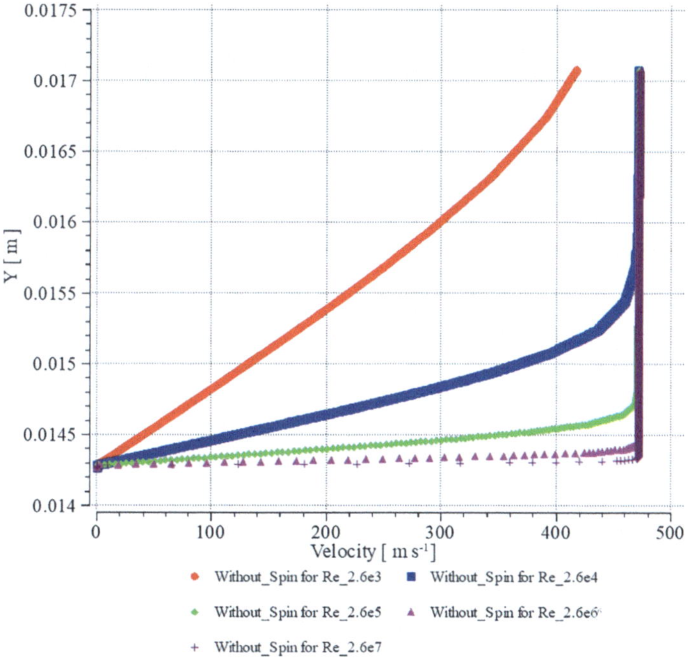

Fig. 9.10 Velocity boundary layer comparison with different Reynolds number

Figure 9.11 shows velocity contour of constant spinning cone at different Reynolds number. When spin is imparted, the velocity boundary layer becomes concave towards velocity axis (Fig. 9.12). As the Reynolds number decreases this concave part can be seen more easily (Fig. 9.12).

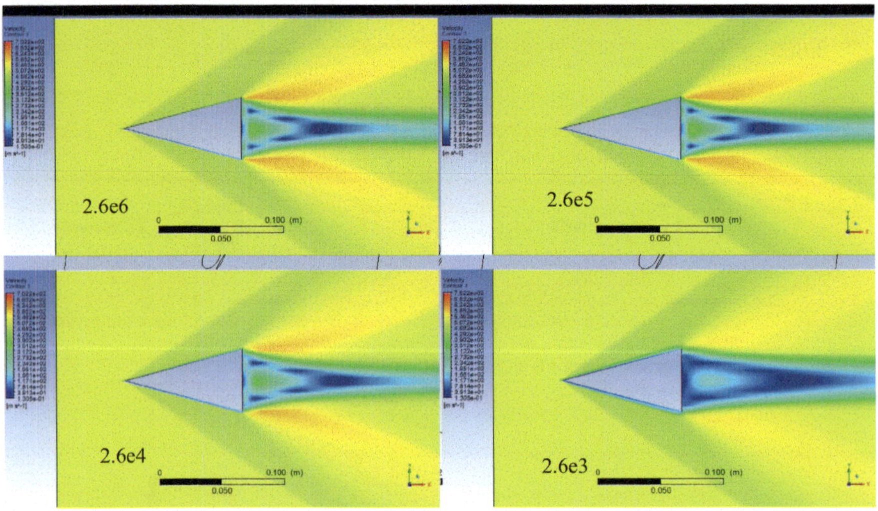

Fig. 9.11 Velocity contour comparison at constant spin (Sr_0.25) with different Reynolds number

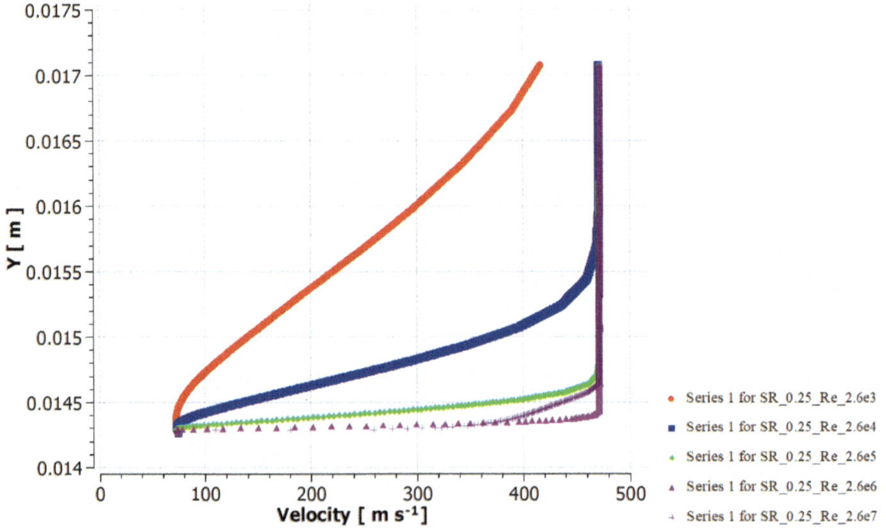

Fig. 9.12 Velocity plot comparison within boundary layer near the wall at constant Sr_0.25 with different Reynolds number

9.3.3 Mach Number Change

The behavior of velocity distribution changes from subsonic regime (Mach less than 1) to supersonic regime (Mach greater than 1). The boundary layer plot for Figs. 9.13 and 9.14 shows the steep gradient in density within very small region (within 0.05 mm) for Mach 2 but for Mach 0.3 the velocity and density gradient are considerable distance (0.5 mm). From the contour and plot we can see there is more variation in density near boundary layer of Mach 0.3 than Mach 2. Thus, the studies for boundary layer experimentally will be easy for Mach 0.3. Thus, schlieren as well as PIV are preferred one to investigate the velocity boundary layer changes.

The effect of spin at different Mach number was also investigated and results are plotted in Fig. 9.15 for velocity distribution and Fig. 9.16 shows the density plot at constant Strouhal number of 0.25.

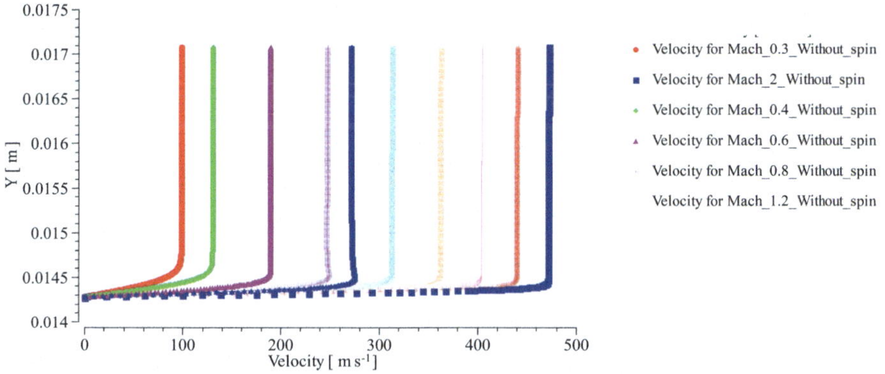

Fig. 9.13 Velocity boundary layer comparison for Mach number change without spin

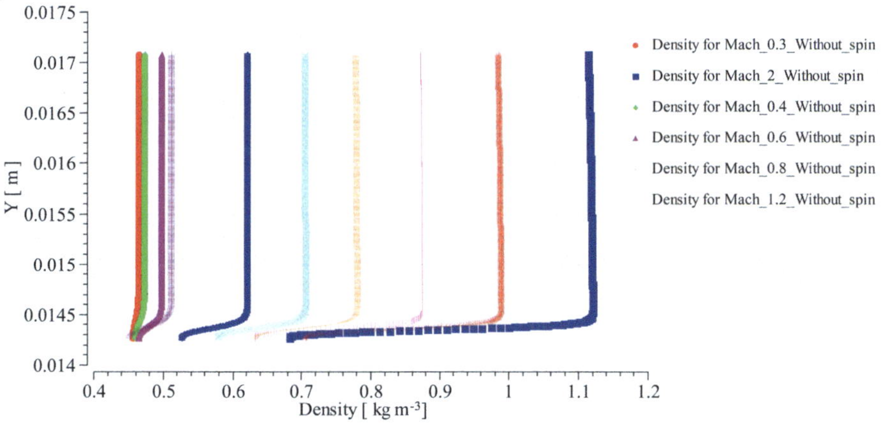

Fig. 9.14 Density variation within boundary layer without spin at different Mach number

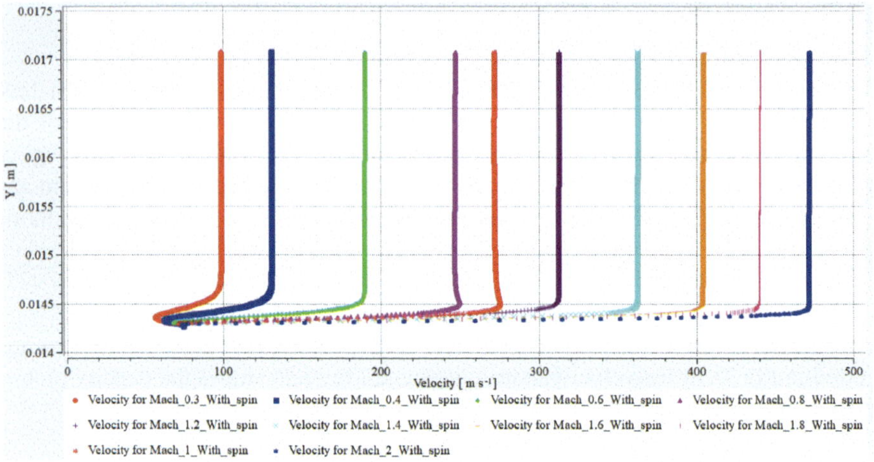

Fig. 9.15 Velocity within boundary layer at Strouhal number 0.25 with varying Mach number

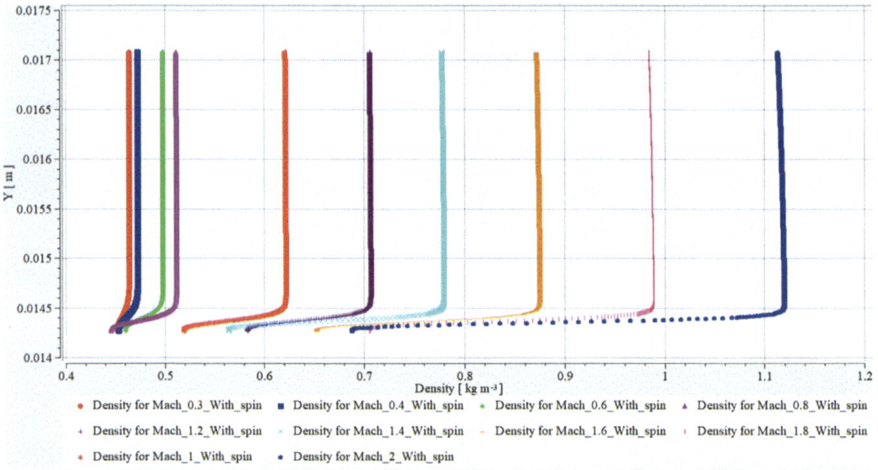

Fig. 9.16 Density boundary layer at Strouhal number 0.25 comparison for different Mach number

It can be seen in Fig. 9.17 that velocity boundary layer comparison is more clearly visible in contour for Mach 0.3 as compared to Mach 2 contour in Fig. 9.18. Figure 9.21 shows the comparison of velocity contour between zero Strouhal number and 0.25 Strouhal number at Mach 0.3 flow.

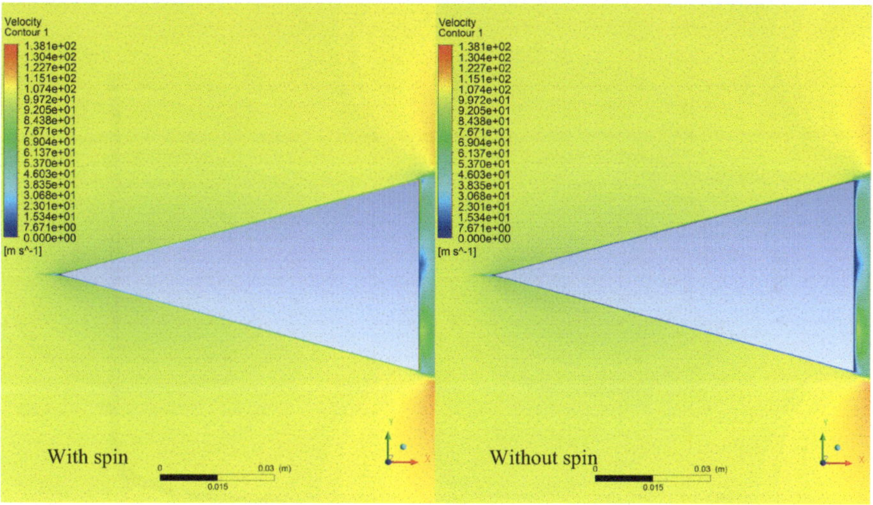

Fig. 9.17 Velocity contour comparison at Mach 0.3 without spin versus with spin

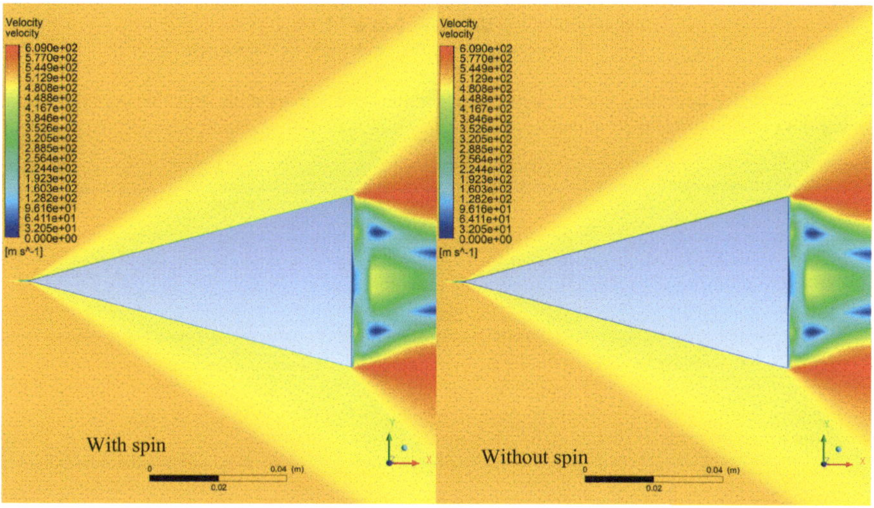

Fig. 9.18 Velocity boundary layer contour comparison without spin versus with spin at Mach 2

Effect of spin on velocity plot are compared for subsonic 0.3 Mach flow in Fig. 9.19 and same plot for Mach 2 are compared in Fig. 9.20. Significant shift in pressure profile is more pronounced by varying spin at lower Mach number of 0.3 as shown in Fig. 9.21 as compared to at Mach 2 in Fig. 9.22.

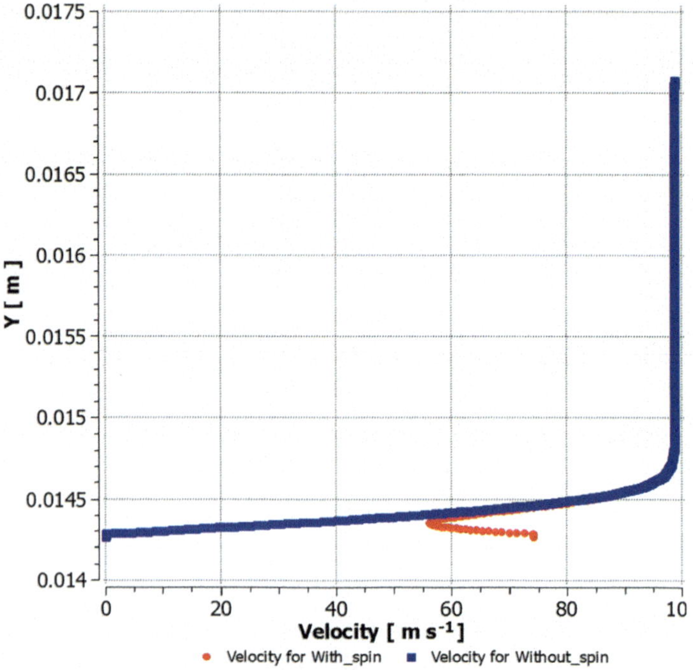

Fig. 9.19 Velocity boundary layer comparison without spin versus with spin (Sr_0.25) at Mach number 0.3

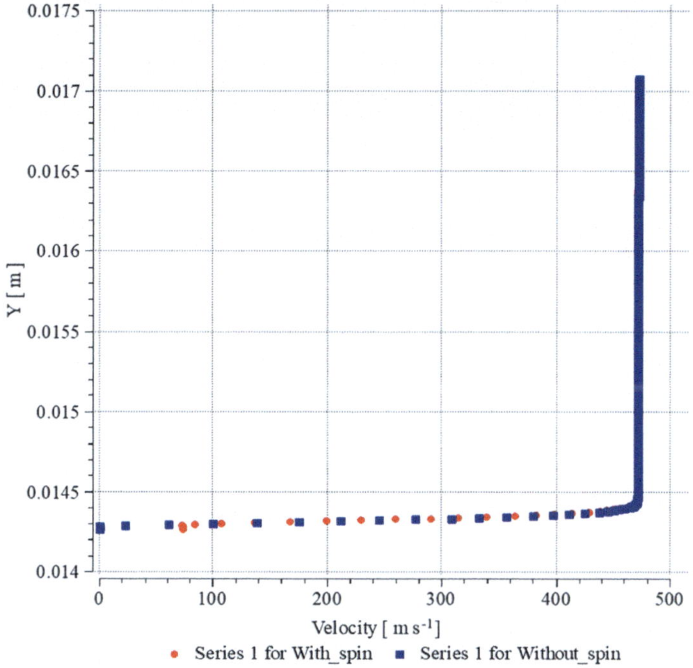

Fig. 9.20 Velocity boundary layer plot comparison without spin versus with spin at Mach 2

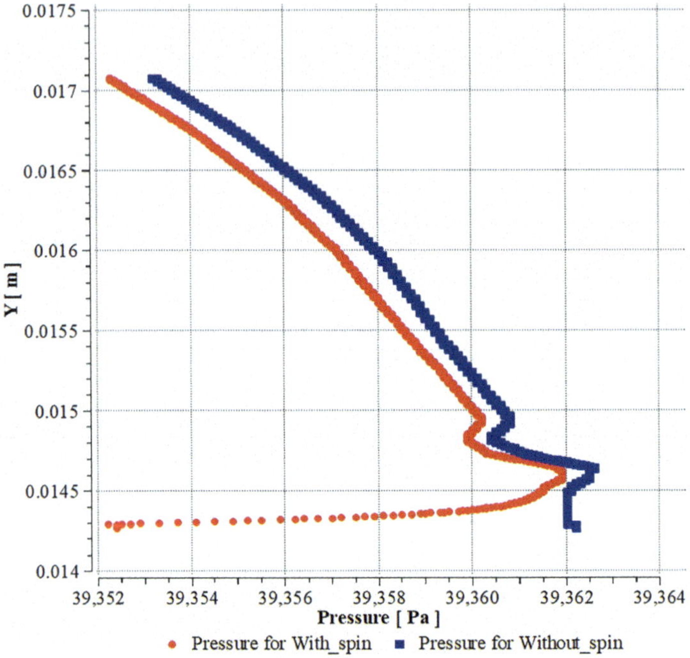

Fig. 9.21 Pressure comparison within boundary layer at Mach 0.3

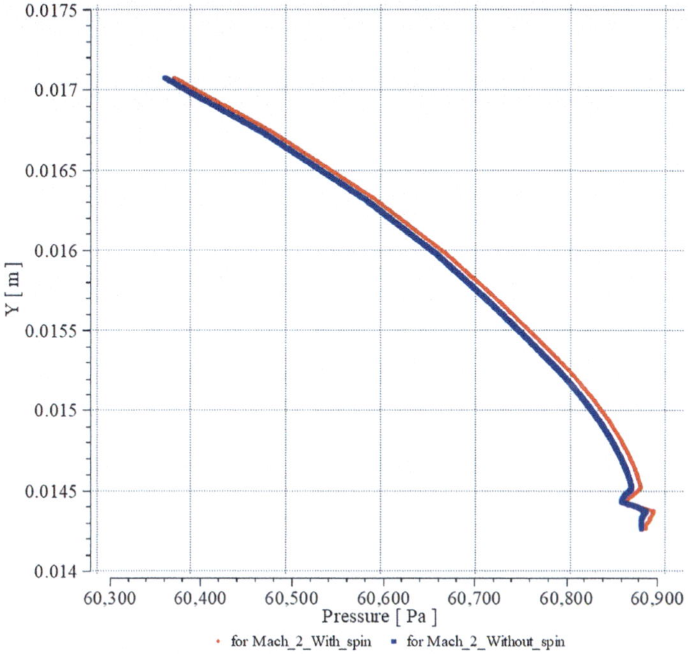

Fig. 9.22 Pressure plot comparison near boundary layer at Mach 2 flow

9.4 Conclusion

The velocity, pressure as well as density near boundary layer of cone surface at location 40 mm from cone base perpendicular to it is compared and plotted by changing the Strouhal number, Reynolds number as well as Mach number. The velocity profile near the wall of spinning cone at lower Mach number is first decreasing from cone surface velocity then it again increases to match up to free stream velocity. Similarly, the pressure at the cone surface is minimum at highest Strouhal number because of highest velocity of cone surface. The This profile is seen more clearly at high Strouhal number and low Mach number. At a constant Mach number when Reynolds number decreases the thickness in boundary layer gets increase. Mach 2 flow had minimum boundary layer thickness among all thus it's difficult to observe the changes using schlieren. Design of experiment should be carefully done keeping these results in mind.

Symbols

$\mathrm{Re} = \frac{\rho \mathrm{VL}}{\mu}$ ρ is density, V is flow velocity, L is cone length and μ is dynamic viscosity.

$\mathrm{M} = \frac{\mathrm{V}}{\sqrt{\gamma \mathrm{RT}}}$ γ is specific heat ratio, T is temperature, R is gas constant

$\mathrm{Sr} = \frac{\omega d}{2\mathrm{V}}$ ω is angular speed, d is cone base diameter

References

1. P. Verberne, S.A. Meguid, Dynamics of precision-guided projectile launch: fluid–structure interaction. Acta Mech. **232**, 1147–1161 (2021)
2. W. Yu, X. Zhang, Aerodynamic analysis of projectile in gun system firing process. J. Appl. Mech. **77**, 051406–1 (2010)
3. M.C. Miller, *Wind Tunnel Measurements of the Magnus Induced Surface Pressures on a Spinning Projectile in the Transonic Speed Regime*, AIAA-83-1838 (1983)
4. D. Klatt, R. Hruschka, F. Leopold, Numerical and experimental investigation of the magnus effect in supersonic flow, in *AIAA Aerodynamics Conference*, 3230 (2012)
5. W.B. Sturek, Boundary-layer studies on spinning bodies of revolution, in BRL Army Ballistic Research Laboratories, Report No. U.S. 1801, Aberdeen Proving Ground, Maryland, AD-785 (1973)
6. M. Pechier et al., A combined theoretical—experimental investigation of magnus effects, in *American Institute of Aeronautics and Astronautics*, A98–32474, P (778–788) (1998)
7. W.B. Sturek, Boundary layer studies on spinning cone, U.S. Army Ballistic Research Laboraties, AIAA No. 72–967.
8. J. D. Anderson, *Modern Compressible Flow: with historical perspective*, vol. 12, McGraw-Hill New York (1990)
9. F. M. White, *Viscous Fluid Flow*, vol. 3, McGraw-Hill New York (2011)

Chapter 10
Experimental Investigation of RR-MR Transition in Curved Shock Wave Reflections

Sushmitha Janakiram and Subrahmanyam Duvvuri

Abstract The transition between regular reflections (RR) and Mach reflections (MR) of shock waves has pertinence in the design of supersonic and hypersonic air intakes. Traditionally, RR-MR transitions in steady supersonic flows have been studied with planar shock waves, mostly motivated by planar compression ramp systems. In recent years, research interest in curved compression ramp geometries has grown due to the promise of improved compression efficiency over their flat counterparts. The leading-edge ramp shock wave along the external compression surface would be curved in such intake designs. The present work focuses on understanding the mechanics of RR-MR transitions in reflections of curved shock waves off a flat surface, a scenario that is typically found in shock-wave systems that form in air intakes. Curved shock wave reflections off a flat plate were studied experimentally at Mach 6 in the Roddam Narasimha Hypersonic Wind Tunnel (RNHWT) at IISc. Wedge angle variation induced hysteresis associated with the interaction between the incident curved shock and the separation shock (generated over the plate) was found to be absent. It was observed that the overall flow deflection angle corresponding to the transition point for a curved ramp was significantly higher than the value reported in the literature for planar ramps.

Nomenclature

M_∞	Freestream Mach number.
θ_{det}	Wedge angle corresponding to detachment criterion.
θ_{vn}	Wedge angle corresponding to von Neumann criterion.
M_1	Mach number downstream to the curved ramp shock wave at the leading edge.
M_2	Mach number at the planar section downstream to the curved portion of the ramp.

S. Janakiram (✉) · S. Duvvuri
Turbulent Shear Flow Physics and Engineering Laboratory, Department of Aerospace Engineering, Indian Institute of Science, Bengaluru 560012, India
e-mail: sushmithaj@iisc.ac.in

© The Author(s), under exclusive license to Springer Nature Singapore Pte Ltd. 2024

G. Rajesh et al. (eds.), *Advances in Shock Interactions*, Springer Proceedings in Physics 408, https://doi.org/10.1007/978-981-97-6099-2_10

θ_D	Overall flow deflection angle.
δ	Total flow deflection caused by the curved ramp at zero angle of attack.
α	Angle of attack of the shock wave generator.

10.1 Introduction

Experimental studies involving shock wave reflections commonly employ planar wedges as shock wave generators. Steady supersonic flow past such a wedge would produce an attached oblique shock if the wedge angle doesn't exceed the maximum angle through which the flow can be deflected for the given freestream Mach number. In an inviscid scenario, when this oblique shock reflects off a flat surface, two prominent shockwave reflection configurations, regular reflection (RR) and Mach reflection (MR), are possible. The transition between these configurations can be induced either by wedge angle variation or by Mach number variation. The two extreme criteria associated with this transition are called the detachment criterion and the von Neumann criterion. For a given freestream Mach number, as the wedge angle θ is varied, an RR is impossible above the detachment limit θ_{det}, and an MR is impossible below the von Neumann limit θ_{vn}. These two conditions enclose the dual solution domain where both the reflection configurations are theoretically possible. Hornung et al. [1] hypothesised the possibility of a hysteresis associated with the RR $\leftrightarrow$ MR transition due to the existence of the dual solution domain. The first experimental demonstration of hysteresis induced by wedge angle variation was offered by Chpoun et al. [2]. It was later shown that three-dimensional edge effects influenced and promoted the hysteresis observed by Chpoun et al. [3–5]. It was also observed that the hysteresis was much more prominent in experiments conducted in an open jet type wind tunnel [6] in comparison with ones conducted in a closed wind tunnel [7]. All of the above mentioned studies involve symmetric shock wave reflection geometries, wherein the viscous effects are mitigated by introducing a symmetry plane that resembles the reflecting plane in an inviscid scenario.

Although past investigations of hysteresis associated with RR-MR transition were motivated by academic interests, in recent years, these studies have proven to be relevant in analysing the design and performance of supersonic air intakes. Some of the geometries investigated in recent studies resemble the compression systems of the high speed vehicles. In recent years, research interest in curved compression ramp geometries has grown due to the promise of improved compression efficiency over their flat counterparts [8]. The leading-edge ramp shock wave along the external compression surface would be curved in such intake designs. Various analytical and numerical studies have been carried out to understand the nature of curved shock waves. Truesdell [9] calculated the jump in the velocity gradient and vorticity function across a curved shock wave in a steady planar flow. Molder introduced the Curved Shock Theory (CST) [10] that relates pressure gradients, streamline curvatures, and vorticity across a curved shock wave. Pant [11] developed an analytical model for the reflection of a curved shock from a flat rigid boundary in an inviscid flow, and further derived the relation between the curvatures of the incident and reflected shock waves. Molder [12] developed an analytical model for the estimation of flow field across singly and doubly curved shock waves that undergo regular reflection and Mach reflection. The theoretically estimated shock wave structures agree well with those observed in experiments and numerical simulations.

The present work focuses on understanding the mechanics of RR-MR transitions in reflections of curved shock waves off a flat surface. While such studies have been performed with planar oblique shock waves, the authors are not aware of any related work with curved shock waves. Tao et al. [13] studied the viscous effects associated with the shock reflection hysteresis analytically, numerically, and experimentally. Xue et al. [14] experimentally studied the incident shock—separation shock interaction focusing mainly on the transition between regular reflection to Mach reflection. Both these studies involved continuously varying the angle of attack of the shock wave generator as the incident planar oblique shock wave interacts with the flat plate positioned below the shock wave generator. The shock wave generator employed in the current study comprises of a curved ramp whose curvature was designed to yield a constant pressure gradient along the compression surface for a Mach 6 inviscid flow at zero angle-of-attack. In Sect. 10.2 the experimental set-up and the shock wave generator design are discussed. In Sect. 10.3 the experimental results associated with the RR-MR transition induced by wedge angle variation are discussed. Section 10.4 presents brief conclusions.

10.2 Experimental Set-Up

The reflection of curved shock waves was studied experimentally in a flow with free-stream Mach number $M_\infty = 6$ in the Roddam Narasimha Hypersonic Wind Tunnel (RNHWT) at IISc; RNHWT is a 0.5 m diameter enclosed free-jet facility that operates in a blow-down-to-vacuum mode [15]. The tunnel was operated at a stagnation pressure of $p_0 = 11.4$ bar and a stagnation temperature of $T_0 = 457$ K. The freestream flow velocity is $u_\infty = 897$ m/s, with corresponding unit Reynolds number of 1.02×10^7 m^{-1} for all the test cases. The flow features were visualised through Schlieren imaging with high temporal resolution, allowing for a detailed study of RR-MR transitions. All data was recorded at a frame rate of 25,000 frames-per-second.

The shock wave generator used in the current study comprises of a two-dimensional curved compression planar wedge (see Fig. 10.1). The contour of the curved surface is designed such that it has a constant pressure gradient along the compression surface for a Mach 6 inviscid flow at zero angle-of-attack. The ramp contains a short planar section upstream of the curved portion with an angle of $\theta_i = 4^o$ at the leading edge. This produces a weak oblique shock, which gets gradually strengthened and curved by subsequent interactions with the compression waves that are produced by the ramp curvature. The curved portion therefore compresses the flow along the surface from a Mach number immediately downstream to the leading edge shock, M_1, to M_2

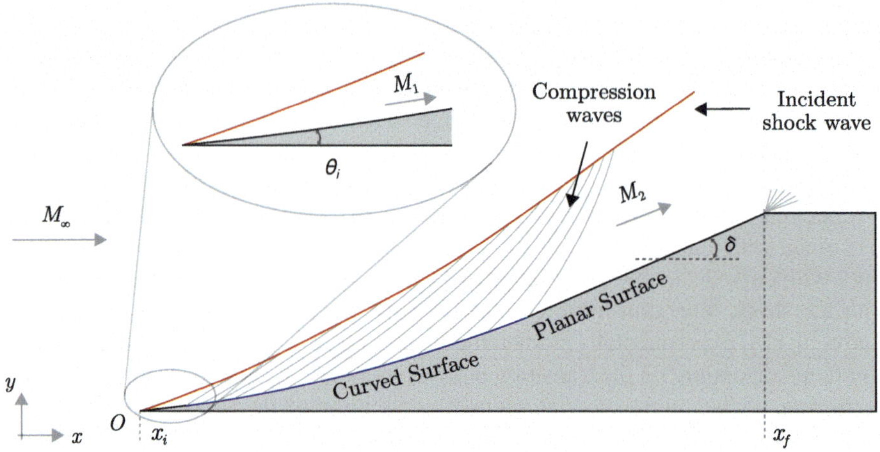

Fig. 10.1 A schematic of the curved ramp

Fig. 10.2 A schematic of the experimental set-up

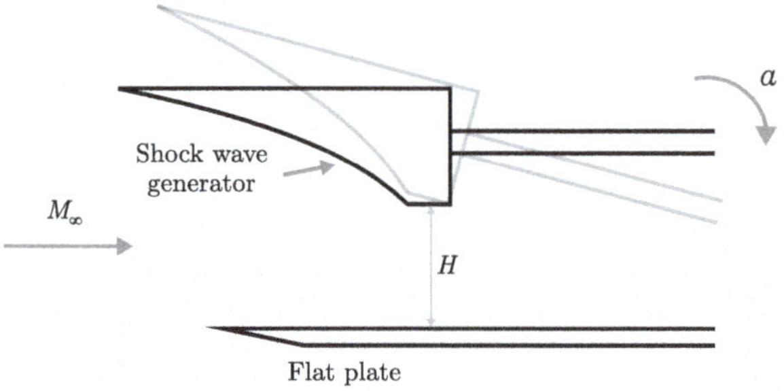

in an isentropic manner; M_2 was set to a value of 4 in this design. The planar section downstream to the curvature is tangential to it and is inclined at $\delta = 19.05°$ with respect to the freestream flow direction. δ also represents the overall angle through which the flow gets deflected when the wedge is at zero angle-of-attack.

The curved wedge was sting-mounted onto a sector arrangement in the RNHWT test section, and its angle of attack was varied continuously during the course of blowdowns by means of a hydraulic piston arrangement (Fig. 10.2). The incident shock from the curved wedge was made to impinge on a flat plate mounted below the shock wave generator. The curved ramp is 100 mm long and 100 mm wide, and the plate mounted below is 160 mm wide. Due to experimental constraints the vertical distance H between the expansion corner of the wedge and the plate was not held constant as the angle of attack of the wedge was varied. Flows with angle of attack increase and decrease were studied in two separate blowdowns. In the first blowdown the angle of attack was increased at a constant rate of 0.16 radians-per-second, and in the subsequent blowdown it was decreased at a constant rate of 0.18 radians-per-second. The set of two blowdowns were performed multiple times to ensure data repeatability.

Fig. 10.3 An instantaneous Schlieren image of Mach 6 flow over the curved compression wedge at zero angle-of-attack. The knife-edge cut is in the vertical direction, and therefore local image intensity represents flow density gradient in the horizontal direction

Figure 10.3 shows an instantaneous schlieren image of the flow, where curved shock wave is easily identifiable. As discussed earlier, a continuous set of compression waves from the curved surface interact with the oblique shock wave generated at the leading edge, causing it to curve by gradually strengthening the shock wave, and thereby resulting in the observed upward concave shape.

10.3 Results and Discussion

Figure 10.4 shows a series of instantaneous schlieren snapshots at three different values of θ_D as the angle of attack is continually increased during the course of a blowdown. Similarly, Fig. 10.5 shows a series of instantaneous schlieren snapshots at three different values of θ_D as the angle of attack is continually decreased during the course of a different blowdown. From these figures it is seen that impingement of the incident shock on the flat plate results in boundary layer separation. The separation point is throughout anchored close to the leading edge of the flat plate. The interaction between the incident shock and separation shock waves is analogous to asymmetric shock wave reflection. With increase in the angle of attack the strength of the incident shock increases and a small, though visible, Mach stem appears at the transition point. As the angle of attack is further increased, the Mach stem height increases and the parting of the two triple points becomes evident. On decreasing the angle of attack, the exactly opposite trend is observed. The Mach stem height decreases and the interaction between the incident shock and the separation shock transitions from Mach reflection to regular reflection. The flow reattachment point moves upstream as the reflection transitions from RR to MR, *i.e.* the streamwise extent of the separation

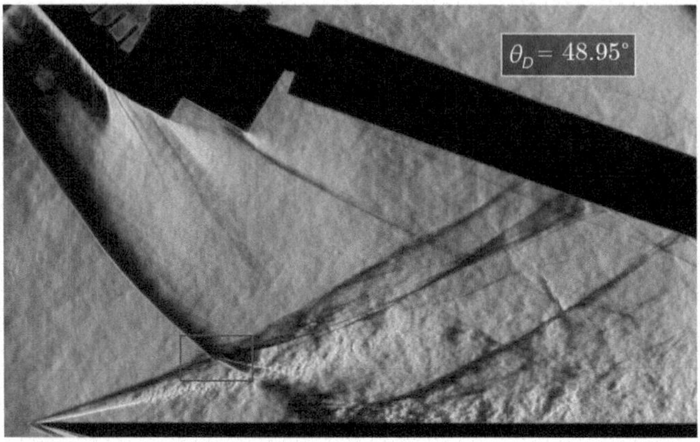

(a) Regular reflection.

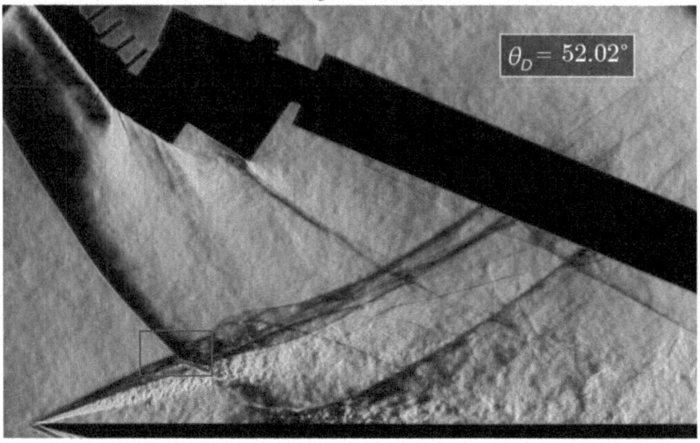

(b) RR-MR transition.

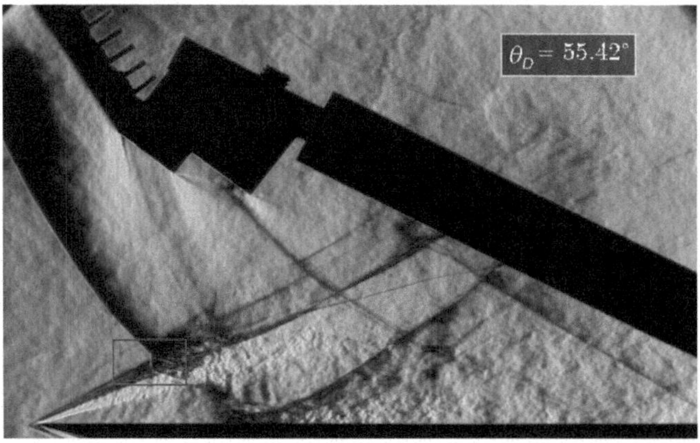

(c) Mach reflection.

Fig. 10.4 Evolution of flowfield as the angle of attack of the shock wave generator is increased. Local image intensity represents flow density gradient in the horizontal direction. The colored box highlights the region of shock interaction

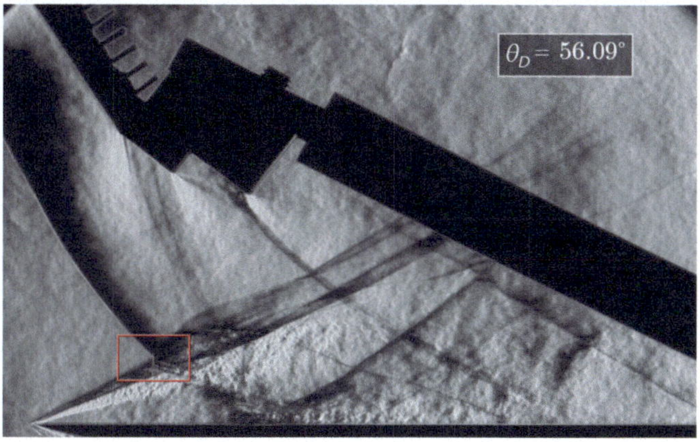

(a) Mach reflection.

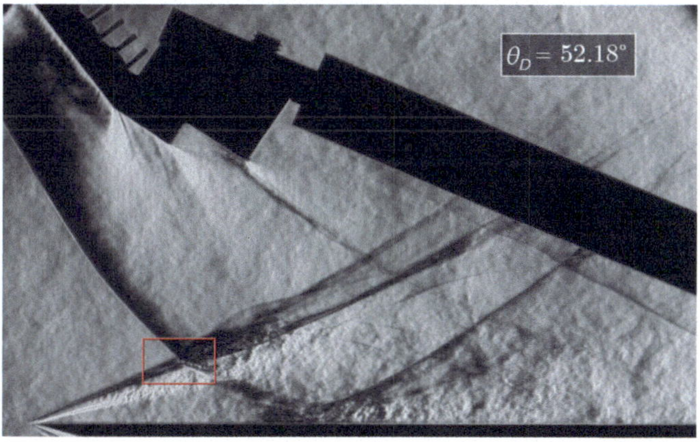

(b) MR-RR transition.

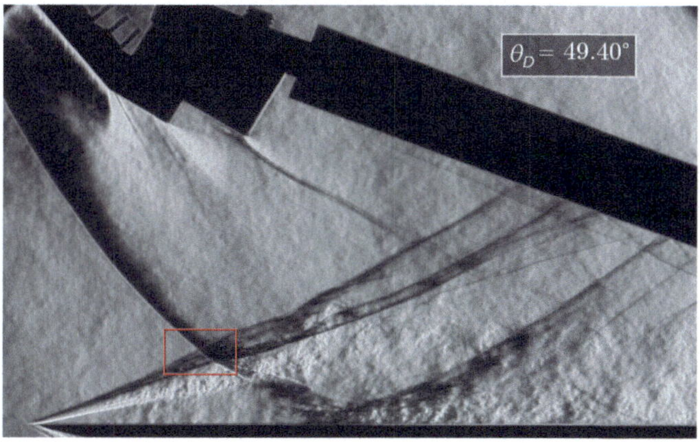

(c) Regular reflection.

Fig. 10.5 Evolution of flowfield as the angle of attack of the shock wave generator is decreased. Local image intensity represents flow density gradient in the horizontal direction. The colored box highlights the region of shock interaction

Table 10.1 Range of angle of attack variation and the transition angle corresponding to the RR-MR transition in terms of the overall deflection angle θ_D

Run	Initial angle (in °)	Final angle (in °)	Transition angle (in °)
1	49.2	55.7	52.1 ± 0.8
2	55.8	49.5	52.2 ± 0.8

bubble decreases. Further, it can be inferred from the Schlieren images that the separation shock wave angle is higher for MR, indicating a stronger shock wave in comparison to RR.

The overall angle through which the freestream flow is deflected by the curved wedge is given by $\theta_D = \delta + \alpha$. Here α is the angle of attack of the wedge with respect to the freestream flow and $\delta = 19.05°$ from earlier discussion. In this paper, the transition angles are reported in terms of θ_D. Table 10.1 shows transition angle corresponding to the two test run cases. The initial and final angles of the wedge for each case are also given; run 1 corresponds to the case where the angle of attack of the wedge was increased and run 2 corresponds to the case where the angle of attack was decreased. It can be seen from Table 10.1 that the RR-MR transition occurs at almost the same angle for both increase and decrease in angle of attack. It is therefore concluded that there is no hysteresis in RR—MR transition in the present experiments. The absence of hysteresis can perhaps be attributed to acoustic waves in the boundary layer, similar to earlier studies with oblique shock reflections off a flat plate [13, 14].

10.4 Brief Conclusions

The current study experimentally investigates the RR-MR transition associated with the reflection of a curved shock off a flat plate. The curved contour of the wedge was designed to hold relevance to ramp geometries used in high-speed air intakes. It was observed that the overall flow deflection angle of $52.0° \pm 0.8°$ corresponding to the transition point for the curved ramp at Mach 6 is significantly higher than the value of $37.8° \pm 0.7°$ reported in the literature for a planar ramp at the same Mach number [14]. A practical implication of these observations is that curved compression ramps would provide a broader window of operation (without transition) in terms of angle of attack in comparison to conventional ramp intakes that induce the same deflection angle. The absence of hysteresis also suggests a more robust operation in the context of air intakes.

Acknowledgements The authors are thankful to B. M. Shiva Shankar, N. Shanta Kumar and M. Harish for their assistance in operations and maintenance of the hypersonic wind tunnel facility. This work was supported by GoI Ministry of Education PhD Fellowship (S.J.) and Saroj Poddar Trust (S.D.).

References

1. H. Hornung, H. Oertel, R. Sandeman, Transition to Mach reflexion of shock waves in steady and pseudosteady flow with and without relaxation. J. Fluid Mech. **90**(3), 541–560 (1979)
2. A. Chpoun, D. Passerel, H. Li, G. Ben-Dor, Reconsideration of oblique shock wave reflections in steady flows. Part 1. experimental investigation. J. Fluid Mech. **301**, 19–35 (1995)
3. B. Skews, S. Vukovic, M. Draxl, Three-dimensional effects in steady flow wave reflection transition. In: *Proceedings of the 12th International Mach Reflection Symposium, Pilanesberg, South Africa*, pp. 152–62 (1996)
4. M. Ivanov, G. Markelov, A. Kudryavtsev, S. Gimelshein, M. Ivanov, G. Markelov, A. Kudryavtsev, S. Gimelshein, Transition between regular and mach reflections of shock waves in steady flows. In *32nd Thermophysics conference*, p. 2511 (1997)
5. M. Ivanov, A. Kharitonov, G. Klemenkov, A. Kudryavtsev, V. Fomin, Influence of test model aspect ratio in experiments on the RR - MR transition, in *The 13th International Mach Reflection Symposium*, p. 68 (1998)
6. V. Fomin, H. Hornung, M. Ivanov, A. Kharitonov, G. Klemenkov, A. Kudryavtsev, A. Pavlov, The study of transition between regular and Mach reflection of shock waves in different wind tunnels, in *Proceeding of 12th Interantional Mach Reflection Symposium Pilanesberg*, South Africa, pp. 137–151 (1996)
7. L. Henderson, A. Lozzi, Experiments on transition of Mach reflexion. J. Fluid Mech. **68**(1), 139–155 (1975)
8. K. Zhang, *Hypersonic Curved Compression Inlet and Its Inverse Design* (Springer, 2020)
9. C. Truesdell, On curved shocks in steady plane flow of an ideal fluid. J. Aeronaut. Sci. **19**(12), 826–828 (1952)
10. S. Mölder, Curved shock theory. Shock Waves **26**(4), 337–353 (2016)
11. J. Pant, Reflection of a curved shock from a straight rigid boundary. Phys. Fluids **14**(3), 534–538 (1971)
12. S. Mölder, Reflection of curved shock waves. Shock Waves **27**(5), 699–720 (2017)
13. Y. Tao, X. Fan, Y. Zhao, Viscous effects of shock reflection hysteresis in steady supersonic flows. J. Fluid Mech. **759**, 134–148 (2014)
14. L. Xue, C. Wang, K. Cheng, A study on the RR-to-MR transition of shock wave reflections near the leading edge in hypersonic flows. J. Fluid Mech. **919**, A40 (2021)
15. V. Sasidharan, S. Duvvuri, Large- and small-amplitude shock-wave oscillations over axisymmetric bodies in high-speed flow. J. Fluid Mech. **913**, R7 (2021)

Chapter 11
Numerical Study of Fabri-Choking Assisted Secondary-Flow Bypass to Enhance the Entrainment of a Supersonic Ejector

P. J. Manoj, C. V. Sumesh Babu, Nikhil Jacob, A. J. Leo, and V. Lijo

Abstract Ejector refrigeration system (ERS) finds extensive industrial applications since it can make use of waster energy. Entrainment ratio (ER), defined as the ratio of mass flow rate of secondary to primary fluid, is the key factor that determines the refrigeration effect of an ERS. The ERS gives maximum ER at double-choked operating condition. That is primary fluid gets choked at the throat of the primary nozzle, whereas the secondary fluid chokes inside the mixing duct at a hypothetical aerodynamic-throat formed by the under-expanded primary-jet. This condition is termed as "Fabri Choking". The mass flow rate of secondary fluid is restricted by the Fabri-choking. This paper proposes a novel geometrical modification, called secondary-flow bypass, that overcomes the mass flow restriction imposed by Fabric choking. To this effect, a novel bypass configuration for the secondary-flow is utilized, which bypasses the location of Fabri-choke inside the mixing duct. Generally, the static pressure inside the mixing duct increases due to pseudo-shocks. However, it is shown that the formation of Fabri-choke is always associated with a sudden reduction in wall static pressure. Present study numerically investigates the flow through modified geometry. As the secondary-flow choked, the series of shocks occurred in the constant-area duct moved downstream from the flow entrainment region, thereby enhancing the mixing and entrainment. In addition, secondary-fluid injection into the primary-nozzle supersonic flow produces complex shock structures inside the mixing duct. Several bypass configurations were numerically analyzed and found the optimal configuration of the bypass for maximum entrainment.

P. J. Manoj (✉) · C. V. Sumesh Babu · N. Jacob · A. J. Leo · V. Lijo
Department of Mechanical Engineering, Government Engineering College Thrissur, Kerala 680009, India
e-mail: manojpjose@gectcr.ac.in

APJ Abdul Kalam Technological University, Thiruvananthapuram, Kerala, India

© The Author(s), under exclusive license to Springer Nature Singapore Pte Ltd. 2024

G. Rajesh et al. (eds.), *Advances in Shock Interactions*, Springer Proceedings in Physics 408, https://doi.org/10.1007/978-981-97-6099-2_11

11.1 Introduction

Supersonic ejectors can function as a compressor, vacuum pumps, or even it can be utilized to mix and transfer fluids and granular materials. The absence of moving parts with zero maintenance, ease of manufacture, and cost-effectiveness are the reasons for the widespread use of supersonic ejectors in many engineering applications.

Ejector refrigeration system (ERS) finds extensive industrial applications since it can make use of waste/solar energy to reduce electricity demand significantly. In ERS, a supersonic ejector replaces the traditional compressor of the vapour-compression refrigeration system. The high-pressure primary fluid enters the supersonic nozzle and is accelerated to supersonic speed. Prevailing low pressure at the exit of the primary nozzle favors suction and entrainment of the secondary fluid from the evaporator. These two fluids mix together inside the mixing duct, which usually involves turbulent mixing, shock boundary layer interactions and complex wall-bounded recirculation patterns. The recompression of mixed fluids to the destination condenser pressure (called back-pressure in this paper) is realized through the pseudo-shocks and flow deceleration in the subsonic-diffuser (Fig. 11.1).

The pumping performance of the ejector can be quantified by the entrainment ratio (ER), defined as ER $= \dot{m}_s/\dot{m}_p$, where $\dot{m}_s$and$\dot{m}_p$ are secondary and primary fluid mass flow rates, respectively. Figure 11.2 shows a typical variation of ER with backpressure (condenser pressure) with two distinct operational modes, namely critical and sub-critical modes. In sub-critical mode, when back pressure exceeds the critical value $\left(p_b > p_b^*\right)$, the secondary fluid becomes unchoked, and *ER* drops drastically (Region I, Fig. 11.2). This is called a single choked state as only the primary fluid is choked. At 'critical backpressure' (p_b^*), , the entrainment ratio is maximum, as both the primary and secondary flows become simultaneously choked (double-choked, Fig. 11.2). Hence, the occurrence of double-choke is the most desired operating condition for the ERS.

Several numerical studies on supersonic ejectors were largely concentrated on shock transformation [1], recirculation bubble patterns [1], mixing characteristics

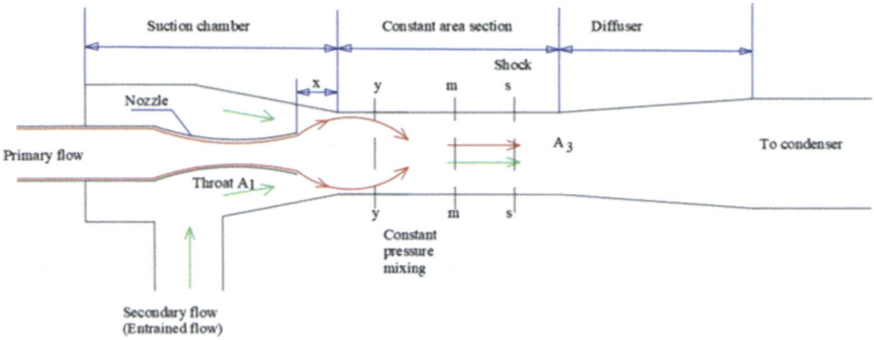

Fig. 11.1 Basic ejector geometry

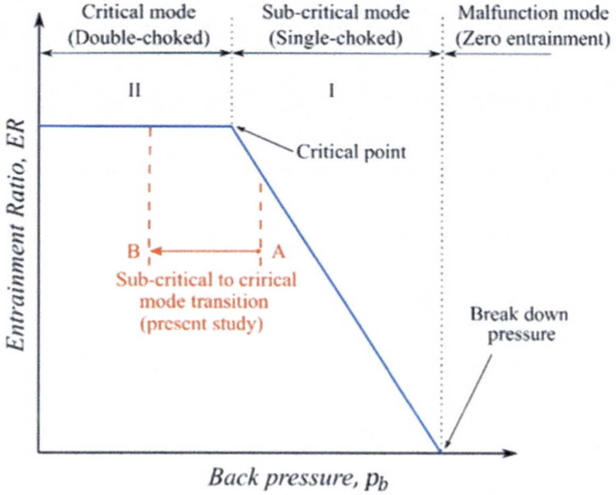

Fig. 11.2 Sketch showing variation of ER with backpressure

[2], selection of turbulence models suitable for ejectors [3]without addressing the mechanism of formation of secondary choke. Lamberts et al. [4] proposed that the location of the Fabri-throat could be ascertained by plotting the sonic line along with the mixing duct wall. Recently Fang et al. [5] numerically visualized Fabri choke (FC) by plotting sonic lines. Arun Kumar and Rajesh. [6] studied shock transformation and hysteresis in under expanded confined jets, Arun Kumar and Rajesh [7] did a study on physics of vacuum generation in zero-secondary flow ejectors, Gupta et al. [2]carried out experimental investigations on mixing characteristics in the critical regime of a low-area ratio supersonic ejector, Bartosiewicz et al. [8] did CFD-experiments integration in the evaluation of six turbulence models for supersonic ejector modeling, Lamberts et al. [9], carried out numerical and experimental evidence of the Fabri-choking in a supersonic ejector. Performance improvement of a bypass ejector with an annular cavity in the nozzle wall was reported by Yinhai Zhu and Peixue Jiang [10]. Such bypass of the primary flow resulted in the reduction of primary mass flow rate, hence improvement in ER. Experimental and numerical validation Yinhai Zhu, Peixue Jiang Bodys et al. [10] discussed a suction bypass implementation which leads to an increase in secondary mass flow rate and hence high ER. Bodys et al. [11], carried out performance improvement of the R744 two-phase ejector with an implemented suction nozzle bypass.

Higher ER denotes higher entrainment capacity of the primary fluid hence greater ejector performance. Several researchers have concentrated on increasing the ER.

The present work concentrated on an ejector operating in the critical mode i.e., with maximum entrainment. A secondary flow bypass is implemented to increase the mass flow rate beyond the mass-flow choking (critical) limit. Fabri-choke always occurs in an ejector operating in the critical mode. During the formation of FC, a low-pressure region is created in the mixing duct. A secondary flow bypass is provided at the location where FC occurs and downstream of it. Flow driving potential for the secondary flow is the low-pressure region created near the FC location. Hence, present study concentrates on the Fabri-choke assisted secondary-flow bypass to enhance ER. To the best of our knowledge, Fabri-choke assisted secondary-flow bypass to increase the secondary mass flow rate beyond the mass-flow choked limit is not reported in the literature.

11.2 Computational Work

The basic dimensions of ejector (Fig. 11.3) are taken from a study on two-dimensional supersonic ejector by Rao and Jagadeesh [12]. The primary flow inlet.

total pressure is 9.89 bar, gauge pressure as 9.87 bar, temperature as 300 K. The secondary flow inlet total pressure as 0.863 bar, and gauge pressure as 0.86 bar. Outlet pressure assumed as 2.39 bar, temperature 300 K. Depth of the ejector 0.08 m. The working fluid is assumed as dry air. Two-dimensional, steady, density-based solver with SST K-w turbulence model is used for analysis.

Secondary-flow bypass is introduced at two locations. For the specified operating conditions two instances of Fabri-chokes were observed. First FC at 70 mm and second FC at 100 mm. All the distances were measured from the secondary nozzle inlet. For the present study, first bypass was provided at exactly at the location of first FC (70 mm) and then bypass was provided between first and second FC (85 mm) Fig. 11.4.

Bypass with constant area, a converging exit and a converging–diverging exit were analyzed. Figure 11.5 compares the experimentally observed variations in wall static pressure with the code for a fixed backpressure (p_b) of 1.34 bar. The ejector was made to operate with an ER of 0.44. Forcefully matching experiment ER resulted in pressure values which are different from the experimentally observed values. However, the general agreement between the predicted wall pressure distribution and the experimental values is good.

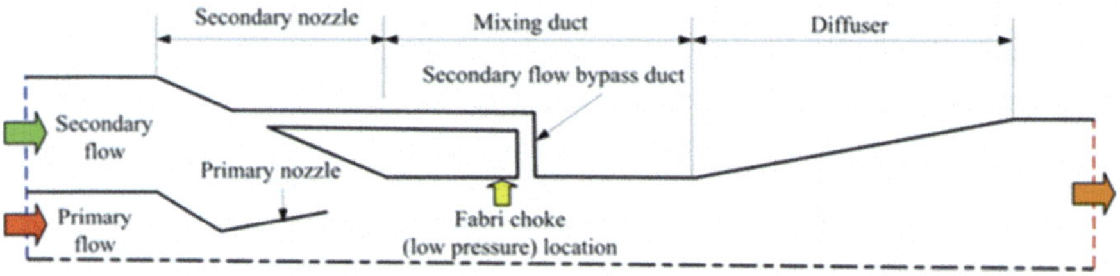

Fig. 11.3 Ejector with secondary-flow bypass

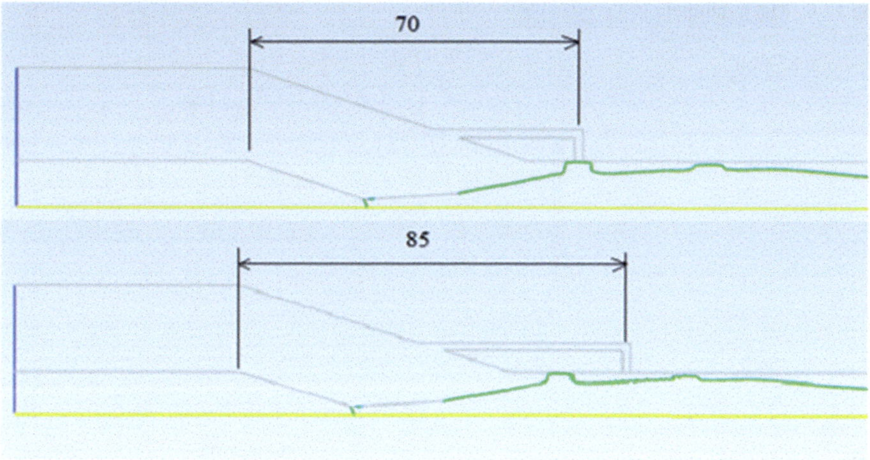

Fig. 11.4 Different locations of bypass

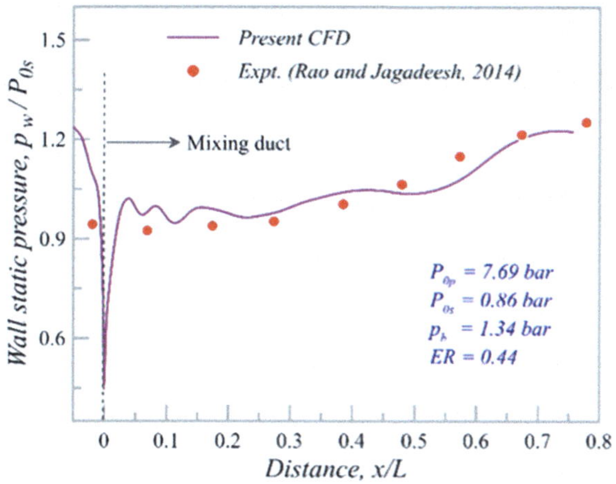

Fig. 11.5 Numerical validation with the experimental results

11.3 Results and Discussion

It was found that secondary-flow bypass has improved the entrainment ratio in all the cases, with a maximum of 7.7% enhancement of ER for a bypass with convergent-divergent nozzle at exit. Mach number variation along the length of the ejector is plotted in Fig. 11.6 for ejector with and without bypass system at location 85 mm. There is a sudden decrease of Mach number at the bypass location, indicating presence of an oblique shock due to secondary mass injection. However, downstream of mass injection, Mach number increases due to the jet expansion and this high Mach region is prevailed a short distance downstream with very weak shocks. The detailed

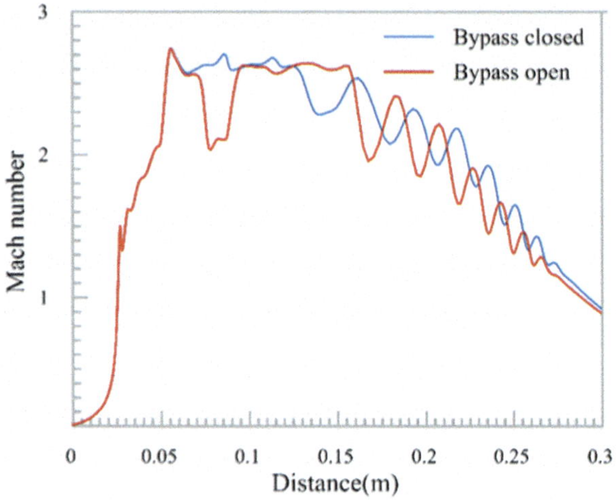

Fig. 11.6 Mach number variation with and without bypass at 85 mm

shock structure is investigated with the help of numerical schlieren in Fig. 11.8. It shows bypass at 70 mm and at 85 mm with a convergent-divergent exit. When FC occurs (with bypass closed), Mach reflection (MR) type of shock reflection can be seen. With bypass open, the MR type of shock reflection changes to regular reflection (RR). Moreover, the supersonic jet expands to fill the entire cross-section of mixing duct, making the flow immune from downstream (backpressure) conditions.

The Mach contours with overlaid sonic lines (to visualize the occurrence of FC) are shown in Fig. 11.7. The lateral development of sonic lines represents secondary flow turning supersonic (FC). The mass flow restriction imposed by FC is removed by secondary mass injection by bypass duct.

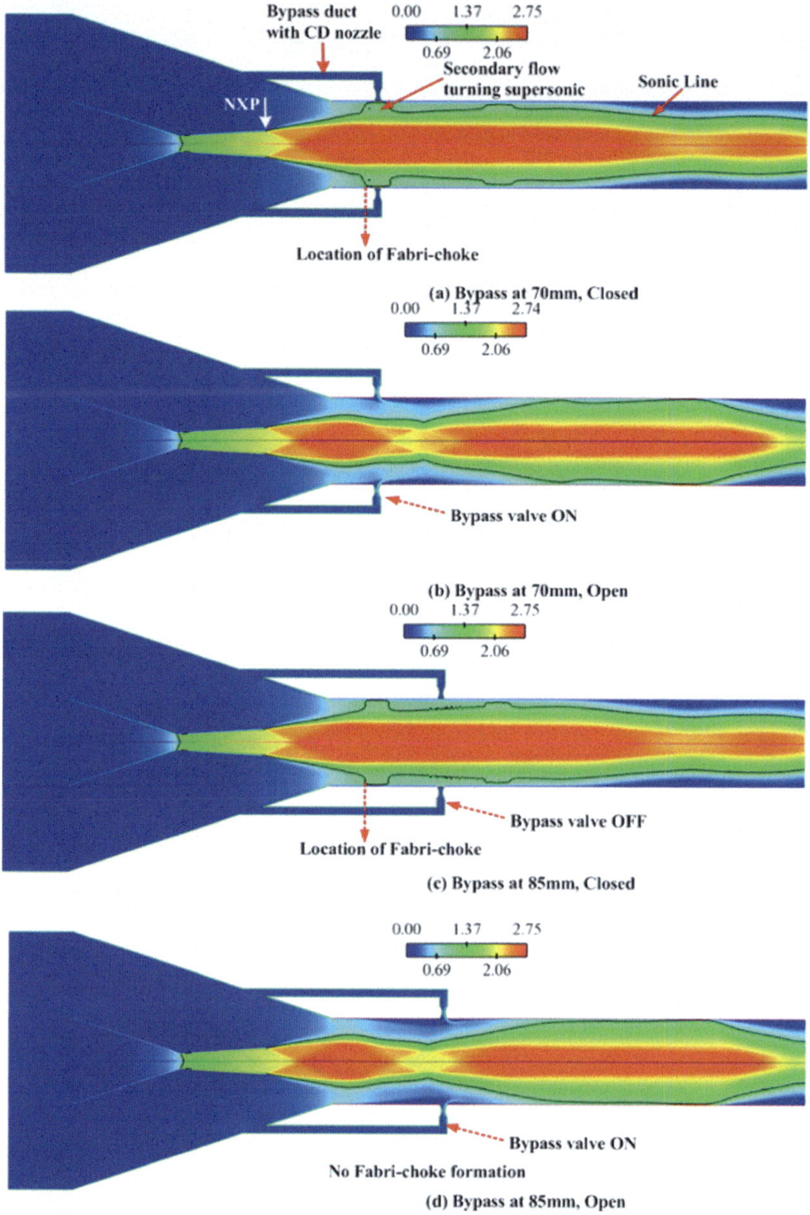

Fig. 11.7. The Mach contours with overlaid sonic lines

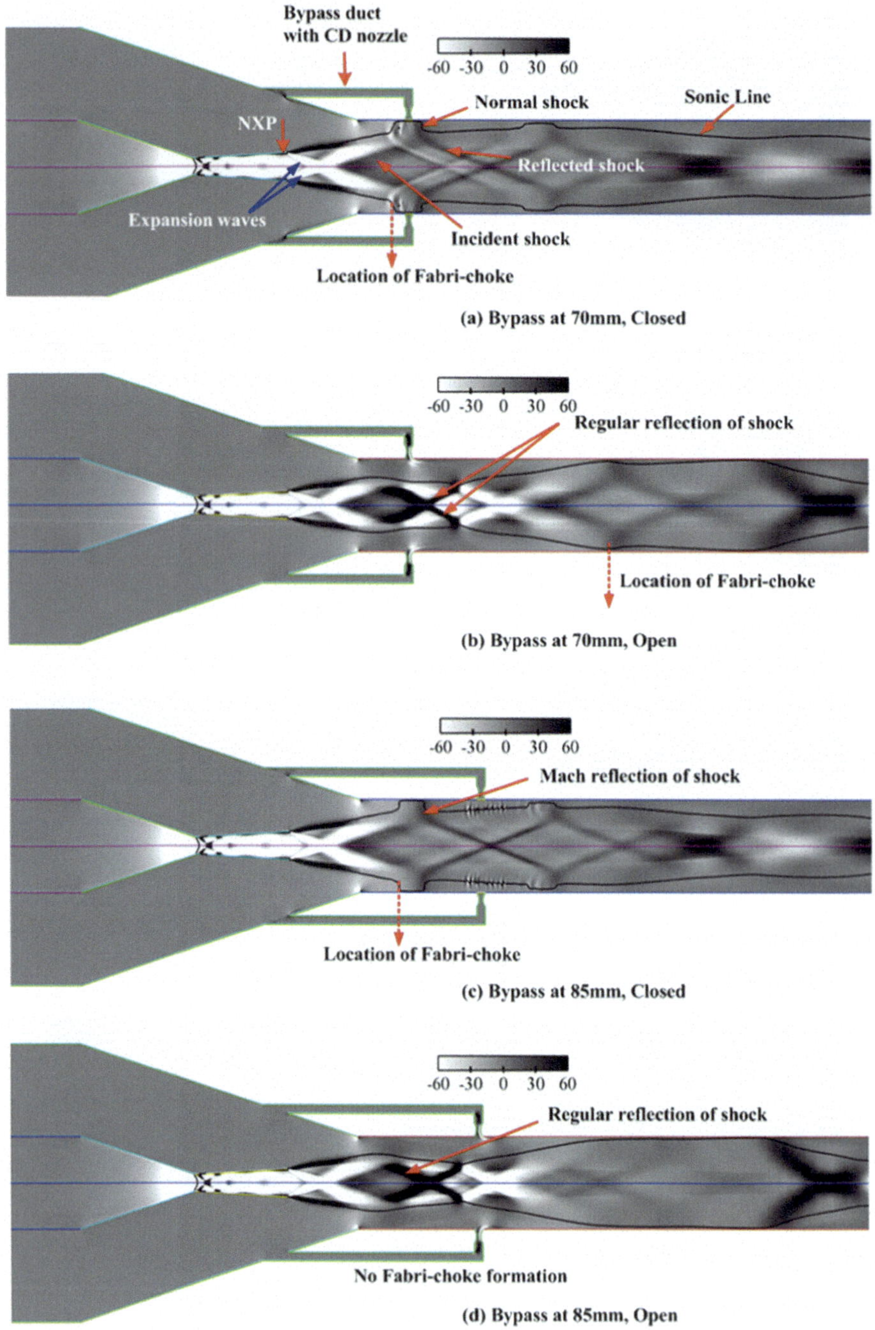

Fig. 11.8 Shock structure with and without bypass

11.4 Conclusion

The present work numerically investigates secondary-fluid injection in a Fabri-choked supersonic ejector typically used in an ejector refrigeration cycle. The occurrence of FC can be considered as ejector operating in the critical mode. Choking of secondary flow imposes a restriction on maximum secondary flow that can be entrained. Present study numerically investigates a method by which mass flow restriction caused by choking of secondary flow, can be eliminated. It was found that there is a sudden decrease in static pressure in the region where FC occurs. The low pressure created due to FC is utilized as a drive the secondary flow using a bypass. Various geometrical configurations were tried for the bypass-duct end geometry. The main objectives of the present effort are to optimize the location of mass injection to maximize ER and the shape of bypass duct end on ER. Numerical simulations are carried out using a 2-D supersonic ejector employing shear stress transport k-ω model with dry air as the working substance. The maximum enhancement in ER is 7.7% with bypass provided at 85 mm (mid-way between first and second FC) with a converging–diverging exit. Also evaluated the effects of secondary mass injection on flow and shock-structures inside the supersonic ejector and found that secondary mass injection forced an MR to RR type of shock transformation, which is not yet reported (Fig. 11.8).

References

1. N. Suvarnakuta, K. Pianthong, T. Sriveerakul, W. Seehanam, Performance analysis of a two-stage ejector in an ejector refrigeration system using computational fluid dynamics. Eng. Appl. Comput. Fluid Mech. **14**(1), 669–682 (2020). https://doi.org/10.1080/19942060.2020.1756913
2. P. Gupta, S.M.V. Rao, P. Kumar, Experimental investigations on mixing characteristics in the critical regime of a low-area ratio supersonic ejector. Phy. Fluid. **31**(2) (2019). https://doi.org/10.1063/1.5078433
3. Z. Aidoun, K. Ameur, M. Falsafioon, M. Badache, Current advances in ejector modeling, experimentation and applications for refrigeration and heat pumps. Part 1: Single-phase ejectors. Inventions. **4**(1). MDPI Multidisciplinary Digital Publishing Institute (2019). https://doi.org/10.3390/inventions4010015
4. O. Lamberts, P. Chatelain, Y. Bartosiewicz, Numerical and experimental evidence of the Fabri-choking in a supersonic ejector. Int. J. Heat Fluid Flow **69**, 194–209 (2018). https://doi.org/10.1016/j.ijheatfluidflow.2018.01.002
5. S. Croquer, Y. Fang, A. Metsue, Y. Bartosiewicz, S. Poncet, Compound-choking theory for supersonic ejectors working with real gas. Energy **227** (2021). https://doi.org/10.1016/j.energy.2021.120396
6. R. Arun Kumar, G. Rajesh, Shock transformation and hysteresis in underexpanded confined jets. J. Fluid Mech. **823**, 538–561 (2017). https://doi.org/10.1017/jfm.2017.231
7. R. Arun Kumar, G. Rajesh, Physics of vacuum generation in zero-secondary flow ejectors. Phys. Fluid. **30**(6) (2018). https://doi.org/10.1063/1.5030073
8. Y. Bartosiewicz, P. Desevaux, Z. Aidoun, Y. Mercadier, *CFD-Experiments Integration in the Evaluation of Six Turbulence Models for Supersonic Ejectors Modeling* (2018)

9. O. Lamberts, P. Chatelain, N. Bourgeois, Y. Bartosiewicz, The compound-choking theory as an explanation of the entrainment limitation in supersonic ejectors. Energy **158**, 524–536 (2018). https://doi.org/10.1016/j.energy.2018.06.036
10. Y. Zhu, P. Jiang, Bypass ejector with an annular cavity in the nozzle wall to increase the entrainment: Experimental and numerical validation. Energy **68**, 174–181 (2014). https://doi.org/10.1016/j.energy.2014.02.046
11. J. Bodys, J. Smolka, K. Banasiak, M. Palacz, M. Haida, A.J. Nowak, Performance improvement of the R744 two-phase ejector with an implemented suction nozzle bypass. Int. J. Refrig. **90**, 216–228 (2018). https://doi.org/10.1016/j.ijrefrig.2018.03.020
12. S.M.V. Rao, G. Jagadeesh, Observations on the non-mixed length and unsteady shock motion in a two-dimensional supersonic ejector. Phys. Fluid. **26**(3) (2014). https://doi.org/10.1063/1.4868879

Chapter 12
Comparative Study on Response of Natural and Artificial Granular Media Towards Blast Loading

Suraj Kumar Mondal, Shweta Paunikar, and G. Jagadeesh

Abstract Conventionally, naturally available sand and soil are used to form a protective layer around the critical military structures. In the recent past, the use of manufactured granular media like glass spheres for protective purposes is increasing. The damping of blast waves due to a granular media is primarily studied from the data obtained from full-scale field experiments. The focus of the present study is to develop a blueprint of an intermediate media between a structure of importance and the region of blast wave impact so that there is minimum or no damage to the structure using shock tubes. At the Laboratory for Hypersonic and Shockwave Research, Indian Institute of Science, to study the effect of a protective layer on the propagation of blast waves, two different granular materials, namely naturally available river sand and artificially manufactured glass spheres of the same particle size, are being used. The Vertical Shock Tube facility is used to create blast wave pressure profiles analogous to free field explosions, and blast waves of different intensities are imparted on the granular media held in a test chamber. For similar packing densities, pressure profiles from the experiments showed that the blast wave through the sand decays faster as compared to the glass spheres. This may be due to the absence of particle interlocking, which is seen prominently in the sand, on account of its uniform spherical shape. Additionally, experiments on glass spheres showed some uncharacteristic pressure jumps at random locations.

12.1 Introduction

The shock waves created due to surface, buried, or air blasts induce vibrations and damage structures, particularly military fortifications, built on or underground. The locally available soil or sand has been typically used as a protective layer around various military structures for efficient mitigation of blast loads. Leong et al. [1] showed that the peak pressure, peak particle displacement, peak particle velocity, and

S. K. Mondal (✉) · S. Paunikar · G. Jagadeesh
Department of Aerospace Engineering, Indian Institute of Science, Bangalore 560012, India
e-mail: surajmondal@alum.iisc.ac.in

© The Author(s), under exclusive license to Springer Nature Singapore Pte Ltd. 2024

G. Rajesh et al. (eds.), *Advances in Shock Interactions*, Springer Proceedings in Physics 408, https://doi.org/10.1007/978-981-97-6099-2_12

peak particle acceleration are the most critical ground shock parameters associated with the vibrations induced due to blast loading. The intensity of blast wave traversing through the ground was characterized based on the type of explosion by Hetherington and Smith [2].

Most commonly, large-scale field experiments are performed to estimate the various ground shock parameters. A bulk of such studies report the behaviour of the locally available soil subjected to underground or surface blasts, for example, TM5-855-1 [3] , Drake and Little [4], Brown et al. [5]. The full-scale field experiments are time-consuming and precarious, and it becomes difficult to understand the physics of wave propagation due to the large number of variables involved. Consequently, researchers have been exploring the use of shock tubes for studying the dynamics of different granular media in controlled laboratory environments [6, 7]. Other granular media, for example, glass spheres, crushed quartz, PVC, potash, plastic spheres, polyethylene spheres, steel cylinders, porcelain spheres, claydite sphere,s have been used for studying the damping of waves by Richart et al. [8], Ben-Dor et al. [9], Britan et al. [10].

From the literature, we note that the damping parameters of naturally available soil are site specific. Furthermore, it is feasible to create similar blast profiles repeatedly in a controlled laboratory environment using shock tubes. Therefore, in this paper, we study the propagation and damping of blast wave profiles generated using vertical shock tubes upon interaction with river sand and glass spheres. The experimental setup is described in Sect. 12.2, followed by the experimental procedure in Sect. 12.4. The details of the experiments performed, the data obtained, and the corresponding observations are discussed in Sect. 12.5.

12.2 Experimental Setup

12.2.1 Vertical Shock Tube

The vertical shock tube (VST) setup designed by Samuelraj and Jagadeesh [11] at the Laboratory for Hypersonic and Shockwave Research, Indian Institute of Science (IISc), Bangalore, used in this study is shown in Fig. 12.1. The driver and driven section of the shock tube are separated by a grooved aluminium diaphragm. The VST comprises of 500 mm long cylindrical sections of SS304 with an internal diameter of 136 mm. The sections can withstand a maximum pressure of 100 bar. The minimum driver length is 500 mm, and the lengths of the driver and driven sections can be adjusted by adding or removing the 500 mm long segments. The ability of the VST to reproduce the features of a near-field air-blast loading like elastic spring back, impulsive loading of a plate, deformation of plates along with numerical analysis are discussed in detail in Samurelraj [12].

Fig. 12.1 Vertical shock tube facility at Laboratory for Hypersonic and Shockwave Research at the Indian Institute of Science, Bangalore

12.2.2 *Test Chamber*

A 770 mm long chamber is designed and used for studying the propagation of the blast wave through a column of granular media. The CAD drawings of the test chamber, along with the sensor locations, are shown in Figs. 12.2 and 12.3. The section made of SS 304 weighs 140 kg and has a flange of an outer diameter of 353 mm, which holds the bolts used to connect the test section to the end of the driven section of the VST. The actual test area is a 684 mm long cuboid with a cross-section of 135 mm × 135 mm. There are 7 ports each on two sides (a total of 14) of the test area with a pitch of 75 mm to hold pressure sensors and as well as accelerometers. Additionally, each of the other two sides of the test chamber is equipped with a window of 230 mm × 60 mm for flow visualization studies.

12.2.3 *Instrumentation*

The sensor ports available in the test section have an outer diameter of 18 mm, and are capable of holding pressure sensors and accelerometers. In this study, we have

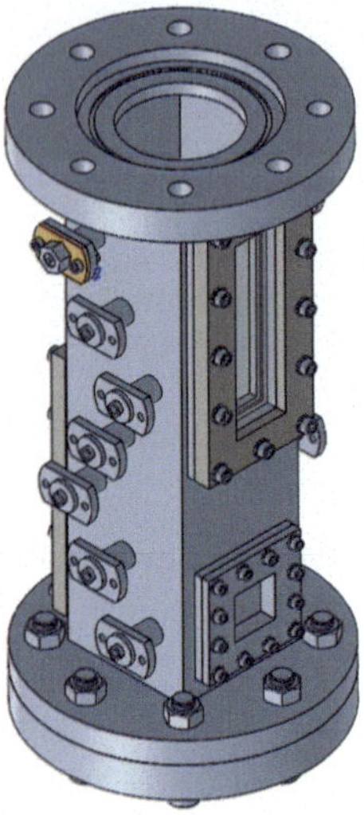

Fig. 12.2 Section for testing the granular media samples

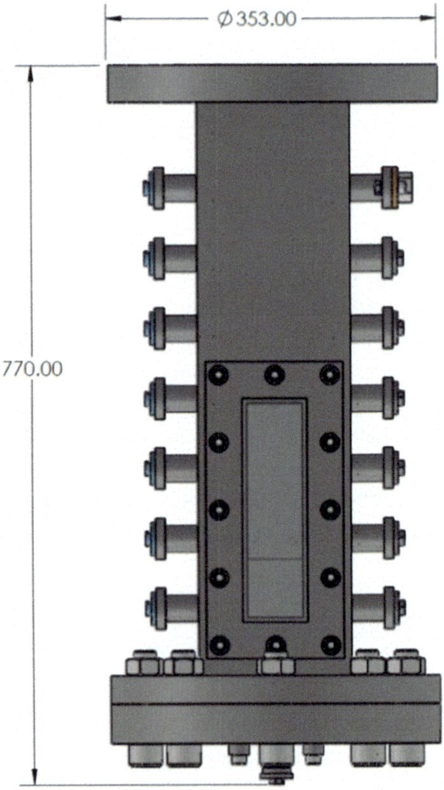

Fig. 12.3 Side view of the test section showing the staggered arrangement of the sensor ports

only used the pressure sensor - PCB 113B24 in a flush mount fashion for side-on dynamic pressure measurement. The sensor weighs 6 g and has a measurement range of 1000 psi with a sensitivity of 5.0 mV/psi (±10%). PCB signal conditioners (482C series) are used to convert the input voltage of the sensors into a readable form, and the Yokogawa DL750 oscilloscope is used for the acquisition of signals at 1 MHz.

12.3 Material Used

River sand abundantly available in IISc is used in this study. To ensure that it is moisture free, the sand is completely oven dried at 105°C for 24 h, followed by air drying for 12 h prior to using it for the tests. The sand has a specific gravity of 2.642 and maximum and minimum dry densities of 1.7134 and 1.495 g/cc, respectively. For the current set of experiments, sand was sieved till 2 mm. In addition to sand, high-performance colourless glass spheres were used for experiments. The particle size of glass spheres lies in the range of 1.5-2.0 mm. As the bulk density achieved by glass spheres was similar to the density of sand for 43 % relative density, the pluviation technique is used to achieve the desired relative density of 43 % for sand.

12.4 Experiments Procedure

The driver and driven section lengths are 0.5 m and 4 m, respectively. The driven section contains air at atmospheric pressure, and helium is used as the driver gas. For the comparative study about the effect of different granular media on blast loading, blast waves of different intensities were imparted onto the granular media column there were made up of either sand or glass spheres. The height of the granular media column in the test section is about 412 mm. A total of 6 pressure sensors are used, out of which 2 are placed in the air column above the free surface of the granular media column, and the remaining 4 are placed within the granular media column. The distance between each of these sensors is 75 mm since the pitch of the sensor ports in the test section is 75 mm. The sensors in the air are 12 mm and 87 mm above the top surface of the granular media samples. The sensors in the samples are 63 mm (D_1), 138 mm (D_2), 213 mm (D_3), and 288 mm (D_4) below the top surface of the granular media. The pressures at these locations are correspondingly denoted as PD_1, PD_2, PD_3, and PD_4, respectively, and P^* for the sensor in air 12 mm above the top surface of the sample.

12.5 Results and Discussion

Experiments have been performed for the relative density of 43 % for sand and glass spheres with 0.5 m driver length. Figures 12.4a, b represent the variation of pressure with time for pressure sensors placed above and below the free surface of the sand column, respectively, for an analogous reflected blast wave of peak pressure equal to 25.12 bar and an impulse of 0.0035 bar-s. Also, Figs. 12.5a, b show typical pressure profiles measured for pressure sensors placed in air and glass spheres column, respectively, for an analogous reflected blast wave of peak pressure equal to 25.38 bar and an impulse of 0.0034 bar-s.

From a simple visual inspection of Figs. 12.4 and 12.5 for blast waves of similar strength impinged on two different granular media mediums, sand and glass spheres, we note that the blast waves impacted on sand column gradually decay along the

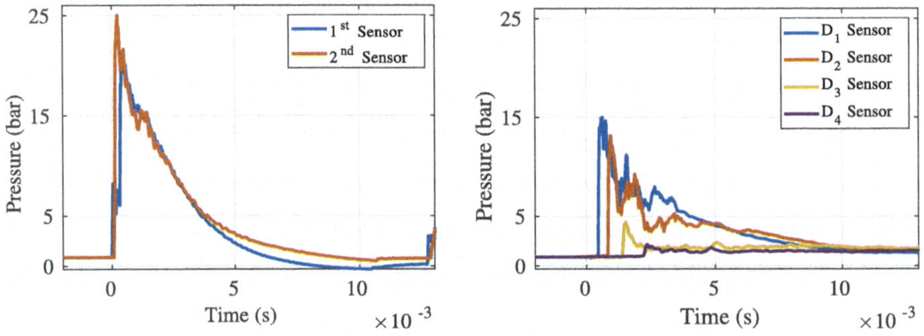

(a) Sensors in air (b) Sensors in sand

Fig. 12.4 Pressure time wave profile recorded at different locations when a sand column is impacted with a blast wave of impulse 0.0035 bar-s

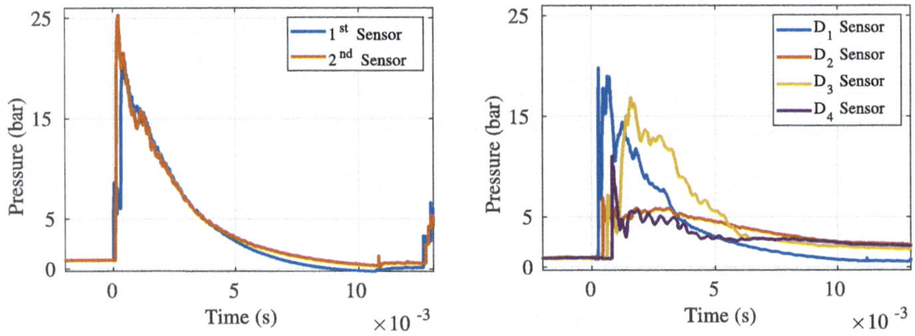

(a) Sensors in air (b) Sensors in glass

Fig. 12.5 Pressure time wave profile recorded at different locations when a column of glass spheres is impacted with a blast wave of impulse 0.0034 bar-s

Table 12.1 Experimental observations for sand

Run No.	Burst Pressure (P_4) (bar)	Impulse (bar-s)	Positive Duration Time (ms)	Mach No.	Reflected P^* (bar)	(PD_1) (bar)	(PD_2) (bar)	(PD_3) (bar)	(PD_4) (bar)
S1	19.51	0.0019	11.17	2.10	12.61	7.68	5.15	2.64	1.32
S2	31.21	0.0027	8.31	2.47	20.50	11.40	6.56	3.66	2.04
S3	42.31	0.0035	8.87	2.61	25.12	15.10	13.23	4.50	2.20

Table 12.2 Experimental observations for glass spheres

Run No.	Burst Pressure (P_4) (bar)	Impulse (bar-s)	Positive Duration Time (ms)	Mach No.	Reflected P^* (bar)	(PD_1) (bar)	(PD_2) (bar)	(PD_3) (bar)	(PD_4) (bar)
G1	22.81	0.0022	10.94	2.08	12.74	6.31	9.31	3.92	3.16
G2	41.71	0.0031	7.55	3.84	20.18	15.56	11.42	10.52	10.32
G3	42.01	0.0034	8.28	2.80	25.38	19.87	6.87	16.96	11.09

length of sand. However, the same observation is not valid for glass spheres media because of uncharacteristic pressure jumps at random locations.

From the experiments conducted to generate blast wave profiles of three different peak overpressures (P^*), namely, 12 bar, 20 bar, and 25bar, in the VST, it was observed that the impulse produced is in the range of 0.0019 bar-s to 0.0035 bar-s. Tables 12.1 and 12.2 contain details of these 3 pairs of experiments on sand and glass spheres for similar blast wave intensities and impulses. In addition to the three most important blast wave parameters (Peak Pressure, Impulse and Decay time), the Tables 12.1 and 12.2 also include the information about the peak pressure attained at each sensor location (i.e., at different depths) inside the granular media column. Figures 12.6a, b, and c present a comparison for attenuation of the strength of pressure waves at each sensor location for 12 bar (Run No. S1 and G1), 20 bar (Run No. S2 and G2), and 25 bar (Run No. S3 and G3) peak reflected pressure, respectively. From these plots, it is evident that stress waves in the sand attenuated quickly. However, when the granular media is made up of glass spheres, there is very little decay in the strength of the stress waves. This can be due to the absence of particle interlocking in case of glass spheres due to their homogeneous shape when compared with sand particles [13]. In order to understand the behaviour of glass spheres in comparison to the sand media upon blast wave impact and the physics of wave propagation, further studies are being conducted with the aid of accelerometers and particle velocity tracking using high-speed imaging.

To understand the physics behind the propagation of high amplitude stress waves in granular media generated due to blast wave impact on it, the average wave velocity at each location is also calculated for each case. From close observation of Tables 12.3 and 12.4 which encapsulate average wave velocity data for each sensor location,

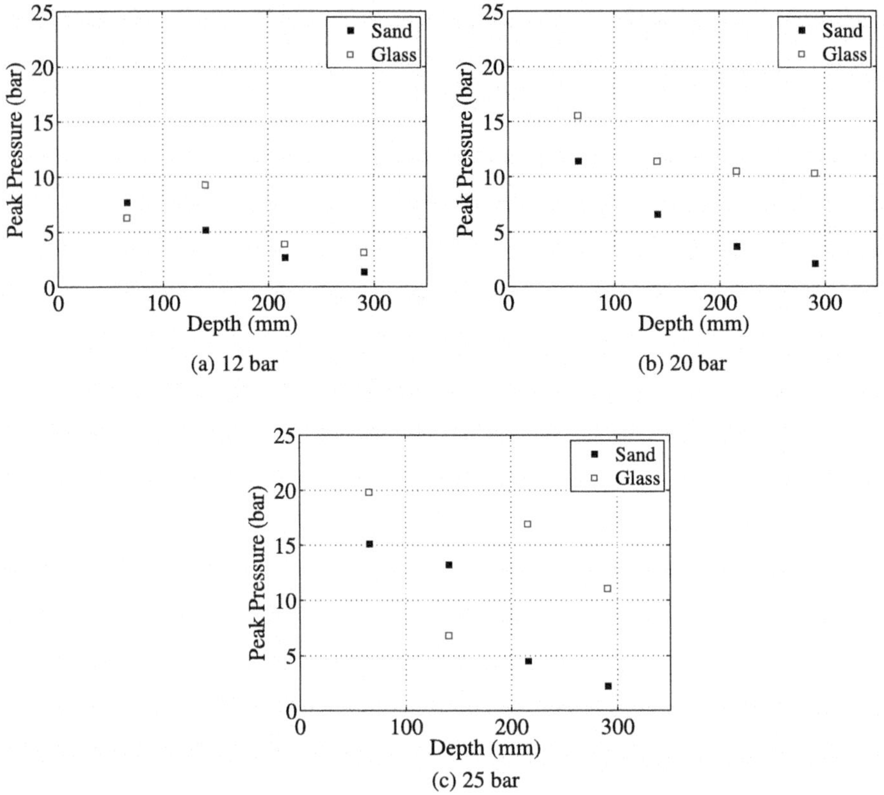

Fig. 12.6 Comparison of the peak pressures observed at various depths along columns of sand and glass spheres for various peak overpressure in air

Table 12.3 Velocities of incoming wave at various depths in the sand sample

Run No.	In air (m/s)	At D_1 (m/s)	At D_2 (m/s)	At D_3 (m/s)	At D_4 (m/s)
S1	727.45	170.27	158.56	123.76	70.09
S2	842.70	169.91	183.37	132.98	104.17
S3	903.61	183.77	204.36	131.12	90.36

Table 12.4 Velocities of incoming wave at various depths in the glass spheres sample

Run No.	In air (m/s)	At D_1 (m/s)	At D_2 (m/s)	At D_3 (m/s)	At D_4 (m/s)
G1	721.15	335.84	414.36	302.42	275.73
G2	1190.47	358.15	446.43	446.43	358.85
G3	974.03	386.51	443.79	419.00	365.85

something exceptional came into view. For almost every run, irrespective of granular media, it was found that in spite of the expected monotonic decay in the wave velocity in granular media, wave velocities in the granular media domain first increased to a maximum value, and after that, waves started to slow down. In the prospect of all complex phenomena happening inside the granular media, the reason we can look up to for this may be the time required for the stress wave to stabilize when it enters the granular media from a gaseous medium. Furthermore, experiments which can help to visualize the actual flow inside the granular media domain may be useful to support this hypothesis.

12.6 Conclusion

Experiments were carried out to understand the propagation of a blast wave generated using a shock tube through two granular media - river sand and glass spheres. From the analysis of these experiments, we can conclude that the attenuation of stress waves is more upon passage through a column of sand than in glass spheres. The attenuation of stress waves observed in the sand may be associated with the particle interlocking. Furthermore, the limited attenuation observed in glass spheres is not uniform. Subsequently, to understand the reason behind uncharacteristic jumps in peak pressures at intermediate depths in the glass spheres media and the reason for an increase in average wave velocity as the wave enters the granular media, further experimentation involving visualization techniques such as Digital Image Correlation is being carried out.

Acknowledgements The research was supported by the project 'Establishing a method for determining the attenuation co-efficient of soil through experiments in a shock tube' funded by DRDO - R & DE (Engrs), Pune, India. The authors would like to thank Machstar group, Bangalore, for their support. The opinions expressed in this paper do not necessarily reflect those of the sponsors.

References

1. E. Leong, S. Anand, H. Cheong, C. Lim, Re-examination of peak stress and scaled distance due to ground shock. Int. J Impact Eng. **34**(9), 1487–1499 (2007)
2. J. Hetherington, P. Smith, *Blast and Ballistic Loading of Structures* (CRC Press, 2014)
3. Fundamentals of protective design for conventional weapons TM5-855-1. US Department of the Army Technical Manual, USA Army Corps of Engineers (1986)
4. J.L. Drake, C.D. Little, Ground shock from penetrating conventional weapons. in *Proceedings of the symposium on the interaction of non-nuclear weapons with structures* (1983)
5. J.L. Brown, T.J. Vogler, L.C. Chhabildas, W.D. Reinhart, T.F. Thornhill, Shock response of dry sand. Sandia Report Sand *3524* (2007)
6. A. Britan, E. Vasil'Ev, I. Zinovik, I.Y. Kamynin, Reflection of a blast-profile shock wave from the end wall of a shock tube. Fluid Dynam. **27**(3), 412–417 (1992)

7. G. Schleyer, M. Lowak, M. Polcyn, G. Langdon, Experimental investigation of blast wall panels under shock pressure loading. Int. J. Impact Eng. **34**(6), 1095–1118 (2007)
8. J. Richart, J. Hall, J. Lysmer et al., *Study of the propagation and dissipation of elastic wave energy in granular soils* (Florida University, Gainesville, Engineering and Industrial Experiment Station, 1962)
9. G. Ben-Dor, A. Britan, T. Elperin, O. Igra, J.P. Jiang, Experimental investigation of the interaction between weak shock waves and granular layers. Exp. Fluids **22**(5), 432–43 (1997)
10. A. Britan, G. Ben-Dor, O. Igra, H. Shapiro, Shock waves attenuation by granular filters. Int. J. Multiphase Flow. **27**(4), 617–34 (2001)
11. I.O. Samuelraj, G. Jagadeesh, Development of a vertical shock tube facility for blast testing applications, in *30th International Symposium on Shock Waves 2* (Springer, 2017), pp. 1333–1337
12. I.O. Samuelraj, Experiments on varying intensity air blasts in shock tubes. Ph.D. thesis, Indian Institute of Science, Bangalore (2019)
13. P. Vivek, Behaviour of granular materials under shock and blast loading-a shock tube investigation Ph.D. thesis, Indian Institute of Science, Bangalore (2018)

Chapter 13
Numerical Analysis on Supersonic Flow Past Partially Covered Cavities

Vaisakh S. Nair and K. Muraleedharan Nair

Abstract The cavity flow studies have been one of the most interesting areas in fluid dynamics, as it involves complex physical phenomenon such as hydro dynamic instability and acoustics response. High speed flow over cavities generates intense pressure fluctuation and compressive waves within the cavity. In the present work, an un-covered and partially covered cavity was considered for the analysis to understand the variations in the fundamental flow behavior when flow experiences a change in the geometry. The partially covered cavities are widely used in landing gear bays, Helmholtz resonator and flame holders in high speed combustion system. The mechanism and the acoustic response of the partially covered cavity is different from that of a baseline uncovered cavity and thus a detailed study is essential for designing an effective flow oscillation suppression technique. The hydrodynamic and acoustic nature of flow around rectangular cavities with and without cover configurations were investigated considering an inflow Mach number of 2.5 and L/D of 7 whereas the Reynolds number based on cavity depth (Re_D) is 45000. The simulations were performed using ANSYS Fluent. The lid on the cavity mouth near the leading and trailing edge affects the overall pressure distribution in and around the cavity. The instabilities associated with the presence of lid along the shear layer near the leading and trailing edge becomes the source for thinning of the shear layer and the production of small scale structures which makes the flow more intricate. The result reveals that, a significant variation in the flow oscillation mechanism occurred due to the presence of lid along the shear layer near the fore and the aft edge of the cavity. The formation of shear layer, compressive waves, their mutual interactions and associated acoustic generation were resolved.

V. S. Nair (✉) · K. Muraleedharan Nair
Department of Mechanical Engineering, Mar Baselios College of Engineering and Technology, Thiruvananthapuram, Kerala 695015, India
e-mail: vaisakh.s@mbcet.ac.in

© The Author(s), under exclusive license to Springer Nature Singapore Pte Ltd. 2024
G. Rajesh et al. (eds.), *Advances in Shock Interactions*, Springer Proceedings in Physics 408, https://doi.org/10.1007/978-981-97-6099-2_13

13.1 Introduction

The cavity flows has got great significance due to its complex behavior such as hydrodynamic instability and acoustics response and fluid wave interactions. High speed flows over cavities find its application in aerospace and defense sectors. The structural vibrations and associated noise generation due to unsteady flow dynamics is one of the inherent features of cavity flows. Based on their length-to-depth ratio (L/D) cavities are classified as deep, open, transitional and closed. It is observed that the flow becomes chaotic as the cavity becomes transitional. The study of cavity flow has always been one of the most intriguing topics among fluid dynamics scholars since it involves various physical attributes, such as aerodynamics and acoustics. Supersonic cavity flows are commonly encountered in high-speed aircraft and defense applications missile bodies.

The static pressure distribution, switching of fundamental frequency and the presence of multiple peaks in the frequency spectra given during the study on varying inflow parameters on the fundamental frequency [1] reveals the influence of inflow conditions on flow oscillation. Another study [2] on transonic flow over cavity illustrates the self-sustaining flow oscillationand acoustic generation. The shear layer, evolves from upstream boundary layer impinges on the cavity's aft wall and produce acoustic wave. These waves reach the cavity's front wall, they would excite and create more instability in the newly developed shear layer at the leading edge. In shear layer mode, the cavity resonant frequencies are dependent mainly on free stream velocity and cavity length and are defined by a semi-empirical relation. An experiment is conducted [3] and found that there is a difference in temperature inside the cavity and the free stream. The author propose a modification on the existing semi-empirical relation and is given as.

$$S_t = \frac{fL}{U\infty} = \frac{(n-\alpha)}{[M\sqrt{}(1+(\frac{r}{2})[\gamma-1]M^2] + (\frac{1}{K})} \quad (13.1)$$

The numerical study [4] using implicit large-eddy simulation for a supersonic cavity flows past an open cavity for an aspect ratio of 2 illustrates the dependence of boundary layer thickness on the oscillation mechanism for an open cavity. The investigation on mixed shear-layer/wake mode in supersonic cavity flows [5] demonstrated movement of compressive waves along with shed vortex towards the downstream and its reflection from the aft wall edge which causes the generation of acoustic wave inside the cavity. The analysis on influence of sidewall, cavity width on flow features [6] explains the confinement of flow passage due to the interaction of wall vortex and the effect of sidewall on flow mechanism. A two-dimensional computation over rectangular cavities of l/d = 6, 12 and 16 for a free-stream Mach number of 1.5 were done [7]. The result shows the formation of the vertical structures from the fore wall. The shear layer separates from the cavity fore wall and reattaches at some point along the cavity floor before reaching the cavity aft wall. The numerical studies [8] focused on distinct features of cavity flow such as generation of compressive wave, shear layer formation, its interaction with the compressive waves etc. Cavity flows are associated with vibrations and noise due to their unsteady fluid dynamics. The investigation on cavity flow is an important research topic involves hydrodynamic and acoustics instabilities [9, 10].

Even though several studies pertaining to flow past cavities are available, limited studies expound the dependence of geometry on flow oscillation mechanism such as flow past partial lid covered cavity structures. The partially covered cavities are widely used in landing gear bays, Helmholtz resonator and flame holders in high speed combustion system. The chamber geometry has an impact on the combustion process and associated variations in the thermo dynamic properties. Effective combustion cannot be initiated until uniform mixing has been achieved. The lid on the cavity mouth near the leading and the trailing edge affects the overall pressure distribution in and around the cavity [11]. The instabilities associated with the presence of lid along the shear layer near the leading and trailing edge becomes the source for thinning of the shear layer and the production of small scale structures which makes the flow more intricate [12, 13]. In the present work we have considered the cavity of L/D 7.0 for an inflow Mach number of 2.5. The study further compares the flow past baseline and partially covered cavities.

13.2 Methodology

Computational domain with imposed boundary conditions is shown in Fig. 13.1. The aspect ratio considered was 7 and the cavity lengthis partially covered with lid cover of $d/D = 0.2$, where d is the of lid thickness. The unsteady, three-dimensional compressible flow is solved in ANSYS FLUENT using k-ω SST turbulence model. A velocity profile is imposed at inlet corresponding to a boundary layer thickness (δ/D) of 0.20, while the wall normal component set to zero, with no slip condition. Neumann boundary condition is imposed at central plane whereas non-reflective boundary conditions are used at inflow and outflow to pass freely the acoustic wave.

A grid independent test was carried out considering four different levels of grid, such as Grid1 ($250 \times 220 \times 32$), Grid2 ($320 \times 300 \times 32$), Grid3 ($350 \times 300 \times 32$) and Grid4 ($400 \times 320 \times 32$). The grid was refined near the walls where velocity gradient was expected to be high as shown in Fig. 13.2, and is allowed to stretch away from the wall.

The non-dimensionalized pressure along the cavity length is shown in Fig. 13.3 for four levels of grid. A noticeable variation in the pressure profile is observed between grid1, grid2 and grid3, whereas no much variation has been observed between grid3 and grid4. This shows that further increasing the grid points not affecting the solution significantly, thus grid 3 has been selected for the subsequent calculations.

The present solver is validated using experimental data [14]. The non-dimensionalized stream wise mean velocity ($U/U\infty$) against normalized cavity length (Y/L) is plotted in Fig. 13.4. The small deviation from the experimental data along the shear layer in the second half of the cavity may be due to the over

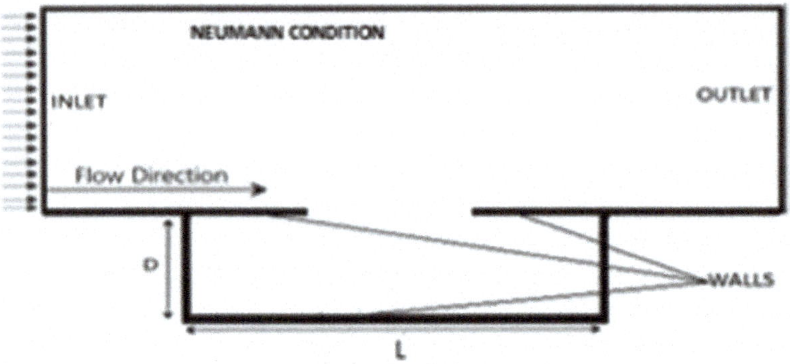

Fig. 13.1 Computational domain with the imposed boundary conditions of partially covered cavity with L/D = 7

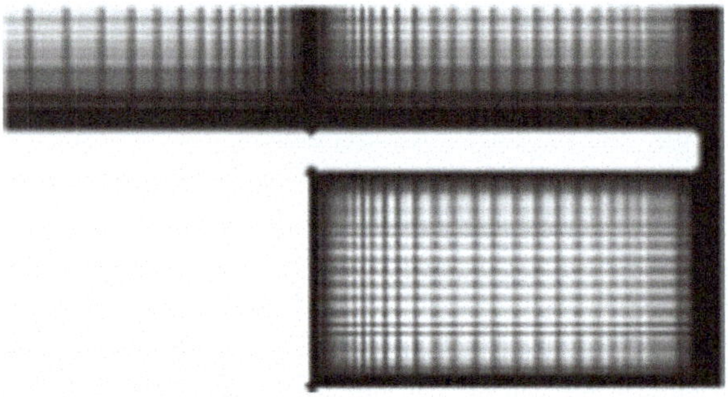

Fig. 13.2 Computational mesh

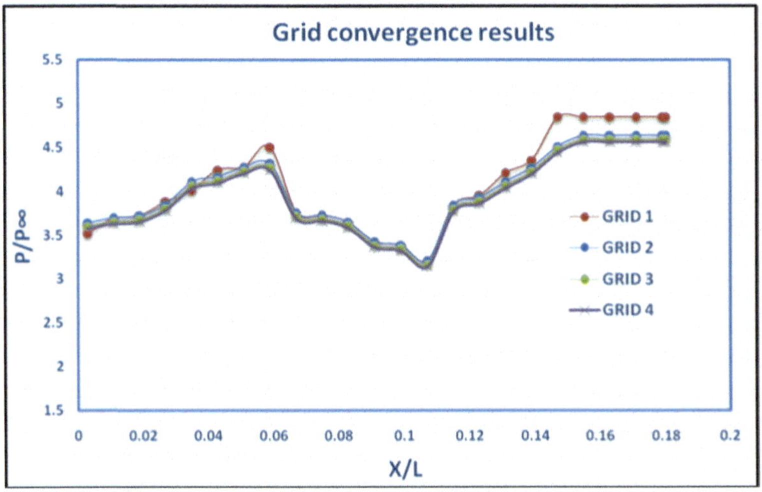

Fig. 13.3 Non-dimensionalized pressure along the cavity length for four different computational grid levels

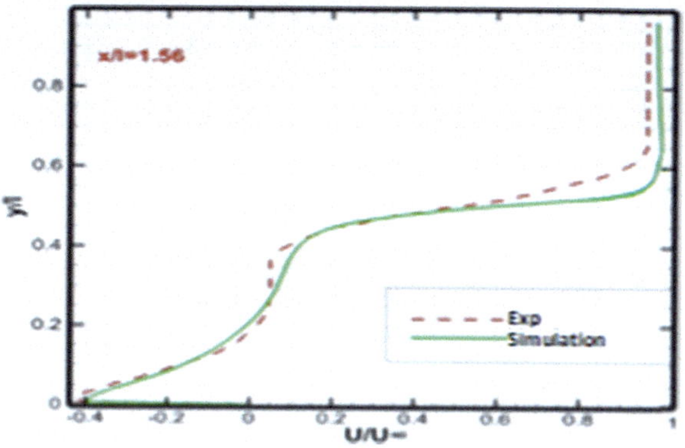

Fig. 13.4 Non-dimensionalized stream wise mean velocity against normalized cavity length

predicted growth rate in that region. However, the trend of the profile obtained from the simulation is inclose agreement with that of experimental data which indicate overall accuracy of the solver.

13.3 Results and Discussion

Supersonic flow of Mach number 2.5 approaches a cavity of aspect ratio of 7.0 with and without lid coverage were considered for the analysis. The fundamental behavior of the fluid flow when encounters a baseline cavity and the changes in the flow features once the lid/partial cover is introduced were observed. The iso-contours of vorticity at two different time instant in a vortex shedding cycle for a partially covered cavity in Fig. 13.5a illustrates the evolution of eddies within the cavity due to the formation of shear layer owing to the existence of velocity gradient between main stream and flow inside the cavity. The formed shear layer moves towards the downstream and partially clipped in to the lid cover near the trailing edge of cavity. The fluid entrained in to the cavity travels upstream and interacts with newly formed shear layer before reaching the fore wall. Here the shear layer turn out to be shorter compared to the baseline cavity. When the flow approaches baseline cavity without cover is shown in Fig. 13.5b. The events such as formation of shear layer, its roll up due to Helmholtz instability and its movement towards cavity bottom before impinging on the aft edge generating pressure pulses inside and outside the cavity. Here the pressure pulse travel upward till the fore wall before it interacts with the incoming shear layer. It is further observed that the recirculation flow from the aft wall is blocked due to the shear layer impingement at the middle of cavity floor. The blockage effect restricts flow to the first half of cavity bottom, which remains as the calm region as depicted in Fig. 13.5b. This phenomenon is absent when the flow past the cavity with cover plate. The compressive waves formed due to flapping of shear layer moves downwards and

(a)

(b)

Fig. 13.5 Contours of vorticity magnitude **a** partially covered cavity and **b** baseline cavity

interact with shear layer during its downward convection. These events contribute significantly to the oscillation mechanism of baseline cavity.

Streamlines of a partially covered and baseline cavity shown in Fig. 13.6 supports the above mentioned vorticity contours behavior of the flow. This illustrates the effect of the partial cover on the baseline cavity.

The Mach contours in Fig. 13.7a, b elucidates the presence of compressive wave within and around the baseline and partial covered cavity and its interaction with the small scale eddies when supersonic flow approaches. When the cavity is partially covered the generated compressive waves from the uncovered portion of the shear layer propagate outward whereas the feedback waves formed within the cavity get reflected from the top cover plate and bottom floor causing an acute acoustic tones formation especially in the second half of the cavity. The pressure lines merged with the iso contours of pressure shown in Fig. 13.8 demonstrates the presence of

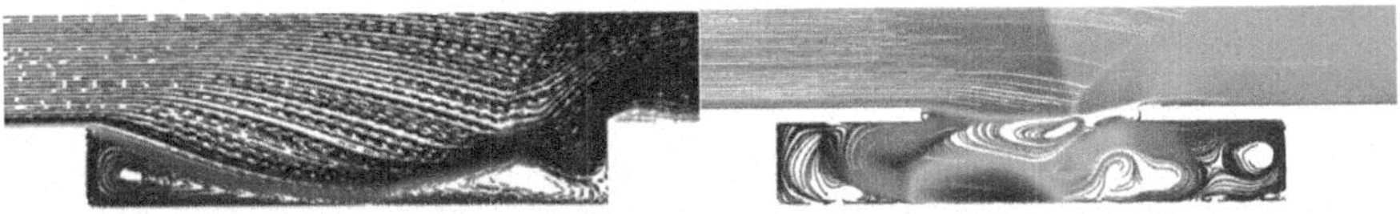

Fig. 13.6 Streamlines-baseline and partially covered cavity

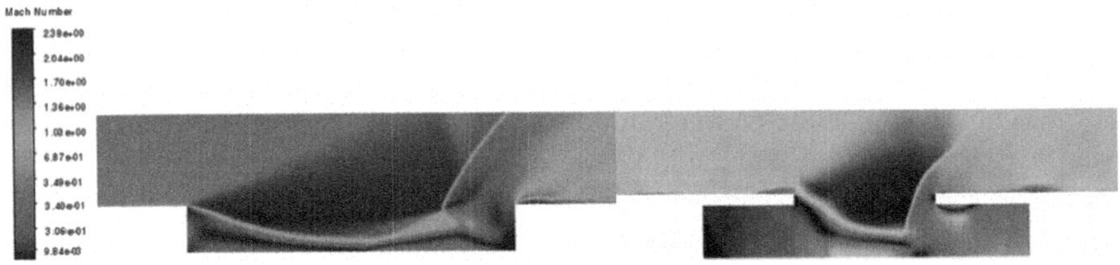

Fig. 13.7 Mach contours for baseline and partially covered cavity

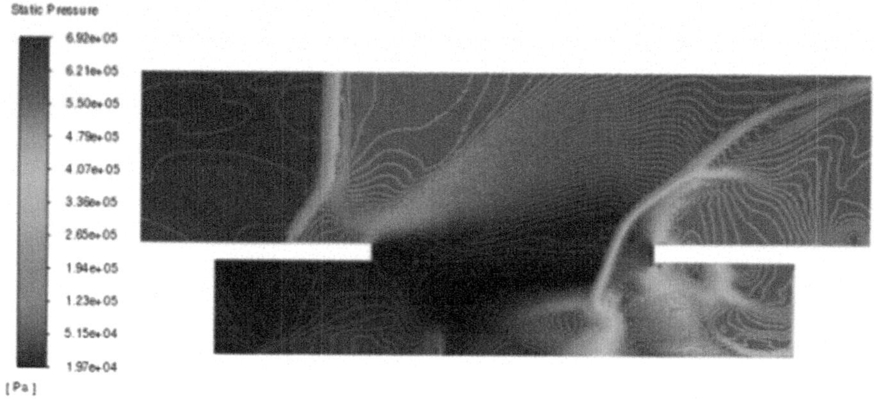

Fig. 13.8 Contours of static pressure for partially covered cavity

compressive waves in the mainstream and within the cavity, their mutual interactions with the shear layer. This compressive wave once formed travel downwards merging together forming lambda shock which influence heavily on flow oscillation mechanism as it travel along with the shear layer.

Pressure signals from the location near the mid of cavity were taken and the sound pressure level spectra were plotted in Fig. 13.9. Multiple frequencies observed along with the fundamental frequency indicate pressure fluctuations within the cavity related to the interaction of recirculation eddies with the cover plate and the bottom wall of the cavity. The deviation in the fundamental frequency form the Rossiter mode is expected because of the presence of cover plates. However fundamental frequencies of baseline cavity are in close match with Rossiter modes. Sound pressure levels are calculated from the pressure signals taken from different location along the cavity mouth and plotted in Fig. 13.10a, b. Due to the fluid acoustic coupling in the uncovered area of the cavity mouth the magnitude of sound pressure levels will be the maximum compared to the covered area. However the magnitude of acoustic disturbance reduced as the flow progress along the downstream side of the cavity at the cavity trailing edge.

The increase in acoustic response within the cavity due to the presence of cover is demonstrated in Fig. 13.11 by considering a stream wise location near aft cover of the cavity and the pressure signals are taken vertically downward up to the cavity floor, The sound pressure levels shows that the acoustic disturbance increases while moving downward towards the floor it may be due to the presence of recirculation exist in that region. The result reveals that, a significant variation in the flow oscillation mechanism occurred due to the presence of lid along the shear layer near the fore and the aft edge of the cavity.

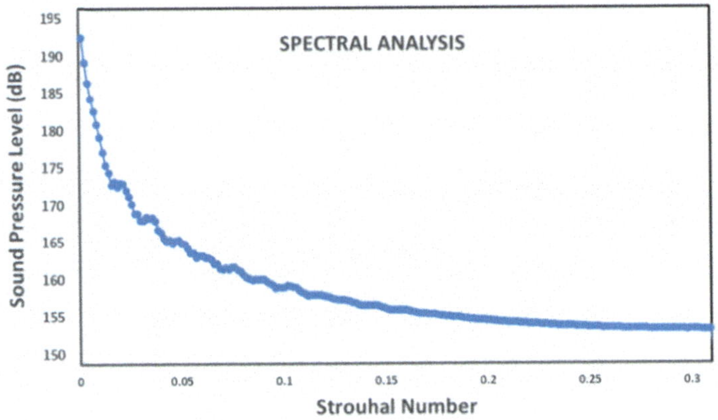

Fig. 13.9 Spectral analysis of the partially covered cavity

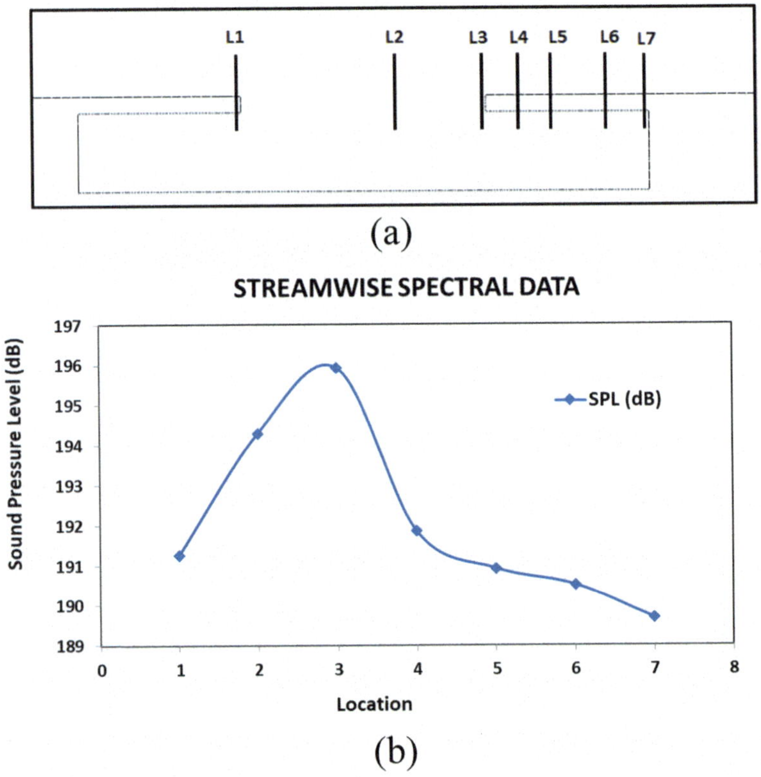

Fig. 13.10 **a** Different stream wise locations along the cavity **b** sound pressure level versus location

Table 13.1 Nomenclature

Symbol	Example	Unit
ρ	Density of air	[kg/m^3]
SPL	Sound pressure level	[dB]
Pa	Pascal	–
d	Thickness of the lid	[m]
D	Depth of cavity	[m]
L	Length of cavity	[m]
AR	Aspect ratio	–
St	Strouhal number	–
$U\infty$	Free stream velocity	[m/s]
f	Frequency	[Hz]
n	Mode number	–
M	Free stream Mach Number	–
r	Recovery factor	–
γ	Specific heat ratio	–
α	Phase delay	–
K	Ratio of vortex convection velocity to free stream velocity	–

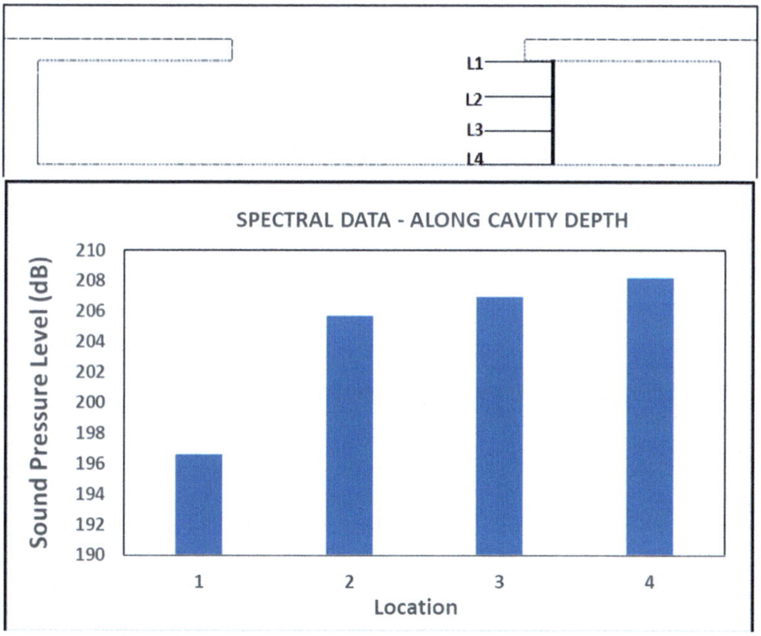

Fig. 13.11 Sound pressure level (dB) versus location—along cavity depth

13.4 Conclusions

In the present work, compressible flow over a partially covered rectangular cavity for an inflow Mach number 2.5 has been analyzed and is compared with the baseline cavity of same inflow conditions. The simulations were performed using ANSYS Fluent. The mechanism of oscillations and associated flow feature was found to be different for the baseline and partially covered cavities. The shear layer leaves the leading edge convects downstream and interact with trailing edge of the cavity. The recirculation eddies travels upstream within the cavity and interacts with the formed shear makes the flow sustained for the base line cavity. The marching of compressive waves that formed from the shear layer and its mutual interaction make the flow complex. The use of partial cover was found to shorten the length of shear layer. The acoustic response of cavity is found to quite different for both the cases. Multiple frequencies observed in the power spectra indicates pressure fluctuations within the cavity which is related to the interaction of recirculation eddies with the cover plate and the bottom wall of the cavity. The fluid acoustic coupling in the uncovered area of the cavity mouth leads to the increase in the magnitude of SPL compared to the covered area. However the magnitude of acoustic disturbance reduced as the flow progress along the downstream side of the cavity trailing edge. The increase in acoustic response within the cavity due to the presence of cover near aft wall of the cavity may be due to the presence of recirculation and their interaction with the compressive waves exist in that region (Table 13.1).

Acknowledgements The authors would like to express their deepest gratitude to computational Research lab, Department of Mechanical engineering, Mar Baselios College of Engineering and Technology for the facilities provided during the course of the work. We would also like to thank the Centre for Engineering Research and Development (CERD) for the support received throughout the work.

References

1. K. Krishnamurty, *Sound Radiation from Surface Cutouts in High Speed Flow*. PhD. Dissertation, California Institute of Technology, Pasadena, California, USA (1956)
2. J.E. Rossiter, *Wind Tunnel Experiments on the Flow over Rectangular Cavities at Subsonic and Transonic Speeds*. Reports and Memoranda No. 3438, Ministry of Aviation, UK (1966)
3. H.H. Heller, D. Holmes, E.E. Covert, Flow–induced pressure oscillations in shallow cavities. J. Sound Vib. **18**, 545–553 (1971)
4. Y. Sun, Y. Zhang, K. Taira, L. Cattafesta, B. George, L. Ukeiley, Width and sidewall effects on high speed cavity flows, in *54th AIAA Aerospace Sciences Meeting*, AIAA Paper, pp. 2016–1343 (2016)
5. S.L. Sridhar, L. Gai, H. Kleine, Oscillatory characteristics of shallow open cavities in supersonic flow. AIAA J. **54**(11), 3495–3508 (2016)
6. S. Pandian, S.L.N. Desikan, N. Sahoo Niranjan, Experimental investigation of starting characteristics and wave propagation from a shallow open cavity and its acoustic emission at supersonic speed. Phys. Fluids **30**, 016104 (2018)

7. O. Baysal, R. Stallings, Computational and experimental investigation of cavity flowfields. AIAA J. **26**(1), 6–7 (1988)
8. C.J. Tam, P.D. Orkwis, P.J. Disimile, Algebraic turbulence model simulations of supersonic open-cavity flow physics. AIAA J. **34**(11), 2255–2260 (1996)
9. T. Colonius, A.J. Basu, C.W. Rowley, Numerical investigation of the flow past a cavity. AIAA paper 99–1912, 10–12 (1999)
10. K. Muraleedharan Nair, S. Vishnu Prasad, S. Vaisakh Nair, K. Greeshma Kartha, Analysis of supersonic flow past deep cavity with different aspect ratios. AIP Conferen. Proceed. **2134**, 020011 (2019)
11. D.J. Wittich, E.J. Jumper, Pressure oscillations resulting from subsonic flow over a partially covered cavity, in *41st Plasma dynamics and Lasers Conference*, Chicago, Illinois, USA (2010)
12. S.A. Syed, K.A. Hoffmann, Numerical investigation of 3-d open cavity with & without cover plates, in *47th AIAA Aerospace Sciences Meeting*, Orlando, Florida, USA (2009)
13. S. Premachandran, Z. Hu, X. Zhang, Computational analysis of partially covered cavity with and without side walls, in *8th AIAA Flow Control Conference*, Washington D. C., USA (2016)
14. N. Forestier, L. Jacquin, P. Geffroy, The mixing layer over a deep cavity at high-subsonic speed. J. Fluid Mech. **475**, 101–145 (2003)

Chapter 14
Microstructural and Texture Investigation on Shock Wave Assisted Deformation of Magnesium

Shreshtha Ranjan, Gopalan Jagadeesh, and Satyam Suwas

Abstract Shock induced deformation of pure polycrystalline Magnesium along perpendicular to basal plane resulted in grain refinement (from 80 μm to 40 μm) mediated through formation of deformation twins (ET). There is a significant weakening of bulk texture has been observed because of grain fragmentation which produces many new orientations and leads to grain refinement. Extension twinning (ET) of the $\{10\,\bar{1}\,2\}$ type at 86° misorientation angle and $<11\,\bar{2}\,0>$ misorientation axis are dominant deformation mechanism in shocked magnesium, also play key role in texture weakening. Mainly, basal (0001) texture weakening was seen, and no other components were observed. Strain gradient within the grain having higher value at the grain boundary whereas peak shift in the kernel average misorientation (KAM) distribution plot by 0.5° (10%) towards higher strain in post deformed sample. Microhardness was found to be highest (50 Hv) at the mid-point deflection region and it decreases systematically towards the periphery.

14.1 Introduction

Dynamic failure analysis of materials is essential in applications such as penetration resistance of defense armors, automotive crashworthiness, the prevention of pressure vessels' catastrophic failure, and aircraft burst containment. The mode of failure of a material varies with its thickness and the projectile's size, shape, and velocity. Therefore, it is critical to understand the complete picture of the material behavior under impact loading to mitigate these risks. Explosive/ Blast forming is also used for high-velocity frictionless forming of thin sheets into complicated shapes. The

S. Ranjan (✉)
Interdisciplinary Centre for Energy Research, Indian Institute of Science, Bangalore, India
e-mail: shreshtharanjan95@gmail.com

G. Jagadeesh
Department of Aerospace Engineering, Indian Institute of Science, Bangalore, India

S. Suwas
Department of Materials Engineering, Indian Institute of Science, Bangalore, India

© The Author(s), under exclusive license to Springer Nature Singapore Pte Ltd. 2024

G. Rajesh et al. (eds.), *Advances in Shock Interactions*, Springer Proceedings in Physics 408, https://doi.org/10.1007/978-981-97-6099-2_14

purpose of the work is to study the microstructural and texture changes under the high strain rate deformation like shock/blast loading for commercial pure magnesium. It is the sixth most abundant element, comprising 2% of the Earth's crust. It has a 1.738 g per cubic centimeter density and is the lightest structural metal known. It has a hexagonal close-packed (HCP) crystal structure with the lattice parameter at 20 °C as a = 0.32 nm, c = 0.52 nm. Magnesium (Mg) alloys are widely used in automotive and aerospace applications due to the combined properties of lightweight, high strength to weight ratio, and high specific stiffness, where they can be exposed to high strain rate situations such as car crashes and ballistic impact.

The deformation mechanism of HCP materials is classified into three groups based on c/a ratio: (a) materials with c/a ratio < 1.63, deformation is accommodated by basal and prismatic slip, examples: Ti, Zr, Be (b) materials with c/a in between 1.63 and 1.73, deformation is accommodated by basal slip and twinning, examples: Mg (c) materials with c/a ratio > 1.73, deformation is accommodated by basal, pyramidal slip and twinning, examples: Zn, Cd [1]. Different slip systems present in the HCP materials are shown in Table 14.1. According to von-misses' criteria, minimum of five independent slip systems are required for plastic deformation. The HCP material processes fewer independent slip systems, which results in twinning to accommodate plastic strain during room temperature deformation. Magnesium is characterized by limited ductility with two independent slip systems, even though many slip systems are presented to the anisotropic slip behavior [2]. In addition to the slip systems, three types of twins can form: (a) extension twins $\{10\bar{1}2\}\langle\bar{1}011\rangle$ with 86° rotation between the twinned and untwined matrix, (b) contraction twins $\{10\bar{1}1\}\langle 10\bar{1}2\rangle$ with 56° rotation between twinned and untwined matrix, (c) double twins $\{10\bar{1}2\}$–$\{10\bar{1}1\}\langle 10\bar{1}2\rangle$ with 38° rotation between twinned and untwined matrix [3]. The critical resolved shear stress (CRSS) for basal slip is nearly 1/100th of the pyramidal slip system. As a result, magnesium and its alloy's activation of the basal slip system are most dominant at room temperature.

Very limited research was done on shock compression of HCP metals. Kanel et al. reported that shock compression of single-crystal magnesium along the c-axis showed an intense amount of twin density compared to other cases [4]. Millet et al. reported that the shear strength of the shock loaded AZ61 magnesium alloy increased

Table 14.1 Slip systems for HCP crystal structure

Type of slip system	Slip systems	Total number of slip systems	Independent slip systems
Basal <a>	$\{0002\}\langle 11\bar{2}0\rangle$	3	2
Prismatic <a>	$\{1\bar{1}00\}\langle 11\bar{2}0\rangle$	3	2
Pyramidal <a>	$\{10\bar{1}1\}\langle 11\bar{2}0\rangle$	6	4
Pyramidal <c + a> -I	$\{10\bar{1}1\}\langle 2\bar{1}\bar{1}3\rangle$	12	12
Pyramidal <c + a> -II	$\{2\bar{1}\bar{1}2\}\langle 2\bar{1}\bar{1}3\rangle$	6	5
Extension twins	$\{10\bar{1}2\}\langle\bar{1}011\rangle$	–	–
Contraction twins	$\{10\bar{1}1\}\langle 10\bar{1}2\rangle$	–	–

with increasing applied shock stress. Also, the shear strength was observed to increase with increasing the time behind the shock front. It was also reported that the major mechanism for deformation was twining followed by dislocation motion at higher stresses [5]. Renganathan et al. reported that the shock compression of single-crystal magnesium along the low symmetric axis is more dominated by the activation of the basal slip. In contrast, the prismatic and pyramidal slip plays a minor role [6]. Asgari reported that when shock loading of WE43 magnesium alloy is done, then a weak basal texture is formed in it along the shock loading direction [7]. H. Asgari reported that as the grain size increases, the intensity of the basal texture increases. It was also reported that the main contribution to deformation in coarser grains was double and contraction twins [8].

Conventional shock tube was used for shock compression of magnesium along c-axis in present experiment. A conventional shock tube is essentially a device in which a normal shock front is produced by the sudden bursting of a diaphragm which separates a gas at high pressure from one at lower pressure. The driven side contains gas at a lower pressure subjected to the shock wave. The high-pressure side contains the driver gas, whose pressure increases until it is enough to rupture the diaphragm. After the diaphragm fails, a compression wave is formed in the driven section, which rapidly steepens to create a shock front. As it travels along the low-pressure section, the shock wave acts like a piston, continuously compressing the gas in the tube [9, 10]. This present work aims shock induced deformation behavior using conventional shock tube to study of microstructural and texture changes during high strain rate impact test in highly pure polycrystalline textured magnesium metal. As-annealed pure textured magnesium was compressed at high strain rate. The deformed samples were sectioned along the rolling direction using Electro-Discharge Machining (EDM) and further characterized for microstructure and texture. Electron back-scattered diffraction (EBSD) studies were done on the transverse plane of the samples while bulk X-ray texture measurement was carried out on the rolling plane.

14.2 Materials and Charecterization

Pure magnesium (Mg) of 5 mm thick was unidirectionally rolled up to 90% at elevated temperature. In each pass of rolling, 10% reduction was given to maintain true strain until sample reduced to 0.5 mm thick sheet, then unidirectionally rolled sheet was annealed at the temperature of 350 ^{O}C for 1 h and circular disc of 8 cm diameter was cut from the rolled sheet by using Electro-Discharge Machining (EDM). The so-obtained disc was subjected to high strain rate deformation normal to the rolling plane in a conventional shock tube. The driven section of the tube was at 760 mmHg pressure whereas the driver section was filled with Nitrogen gas to build up a pressure enough to rupture the diaphragm. After the diaphragm ruptures, a compression wave is formed in the driven section, which rapidly steepens to form a shock front. It is the shock front which deforms the exposed sheet while it is constrained along its

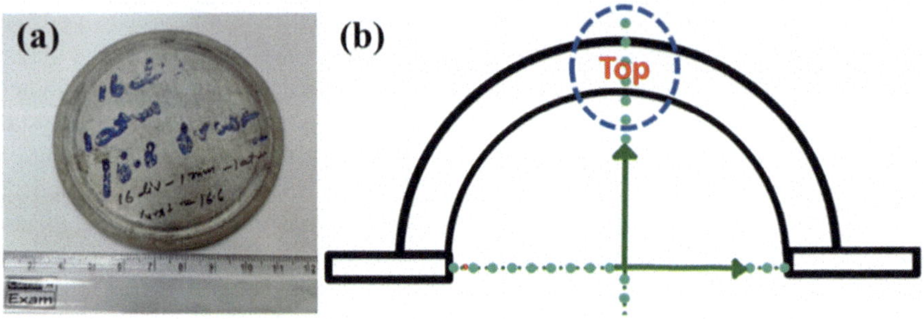

Fig. 14.1 **a** Shocked deformed magnesium sample (hat shape) at 0.8 Mpa of 8 cm diameter **b** schematic of TD plane of sectioned deform sample along RD direction

circular edge. This tends to the formation of a spherical cap from a flat disc as shown in Fig. 14.1.

The shock induced deform magnesium sample was sectioned along the rolling direction using Electro-Discharge Machining (EDM) and further characterized for microstructure and texture. Electron back-scattered diffraction (EBSD) studies were done on the transverse plane of the samples while bulk X-ray texture measurement was carried out on the rolling plane. For EBSD measurements, the transverse surfaces were mechanically polished up to sample roughness of 5 μm (4000 SiC grit paper) followed by manual electropolishing. The electrolyte for the manual electropolishing was prepared by proper mixing of orthophosphoric acid (H_3PO_4) and ethanol (C_2H_5OH) in 3:5 ratio. The 3 V voltage was applied for the first 30 s followed by 1.5 V for the next 120 s. An accelerating voltage was fixed for all EBSD scan was 20 kv. The orientation data obtained from EBSD scan was analyzed using OIM™ Analysis Software. For the bulk texture measurement, the rolling plane of the sheet was polished and investigated in an X-ray texture goniometer based on Schulz reflection geometry using Cu-Kα radiation. Six incomplete pole figures {002}, {100}, {101}, {102}, {103}, and {110} were measured experimentally. Quantitative texture analysis was carried out for calculating the orientation distribution functions (ODF) using resmat software. The hardness of the deformed samples was measured on the transverse section by using the Vickers hardness technique.

14.3 Results and Discussions

Shock pressure of different values are obtained by changing thickness and grooves of diaphragm. 1 mm thickness and 16 divisions grooved Cp aluminium diaphragms is used for shock experiment which was mentioned in Table 14.2. Disc is cut from the rolled specimen respective. That disc is annealed at 350 °C for 1 h to relieve the stress within the rolled specimens. Nitrogen gas was used in driver section and driven tube was fixed at 760 mmhg pressure. 8 bar pressure was recorded at the end of driven tube that is our pressure which felt on magnesium rolled sample on basal plane. Before shock loading the samples, they were polished using SiC emery paper to a grit size of 3000 to ensure that there were no scratches that may act as the crack initiating sites. Strain measurements are done on the shock-deformed specimen using the midpoint deflection method [11]. Maximum deflection was measured by vertical gauge. In table 14.3, effective strain, maximum deflection, radius of curvature and peak pressure are listed. Maximum deflection and strain of shocked magnesium sample are 6.5 mm and 6.52% respectively.

Figure 14.1 shows that (a) shocked deformed magnesium sample looks like hat shape, which was deformed at 0.8 Mpa, and (b) schematic of sectioned shock deformed sample. For EBSD and bulk texture measurement, sample was cut from top of the deformed magnesium that is mentioned in Fig. 14.1a. Maximum strain was experience at where their sample has maximum deflection. Figure 14.2 shows the IPF maps of annealed and shocked samples. EBSD scans of before/annealed and after/shocked specimens which were taken at TD plane, done with a step size of 0.5 μm. It was observed that shock induced deformation of pure polycrystalline magnesium along perpendicular to basal plane resulted in grain refinement (from 80 ± 18 μm to 45 ± 15 μm). Also, there was twin formation after shock deformation. Figure 14.2 clearly shows that annealed sample have no deformation twins but after shock deformation, twins are formed. As it can be observed that grains with different colors are formed apart from red; therefore, it can be inferred that the intensity of the initial basal texture was decreased [12]. This decrease in the basal intensity may be due to the formation and growth of that, which eventually results in the reorientation of the grains.

Figure 14.3 shows the (0002) Pole Figures (PF) for initial sample (annealed) and deformed (Def_1 mm16 div) sample. six pole figures were measured in the texture goniometer. The scan was done till $\psi = 80°$. Thus, the software obtained the remaining part of the pole figure. It is apparent from the above figure that a strong basal texture was formed along the normal direction in annealed sample. There is a significant weakening of bulk texture has been observed in shocked sample because of grain fragmentation which produces many new orientations and leads to grain refinement.

Table 14.2 Driven pressure, gas used in driver section, and peak pressure during experiment

Diaphragm	Driven pressure (mm Hg)	Gas used in driver section	Peak impact pressure (MPa)
1 mm 16 div	760	Nitrogen	0.8

Table 14.3 Effective strain, deflection, and radius of curvature of shock deformed magnesium

Materials	Peak pressure (MPa)	Deflection (d_f) mm	Radius of curvature (ρ) mm	Strain (ε)
Magnesium	0.8	6.5	51.32	6.52

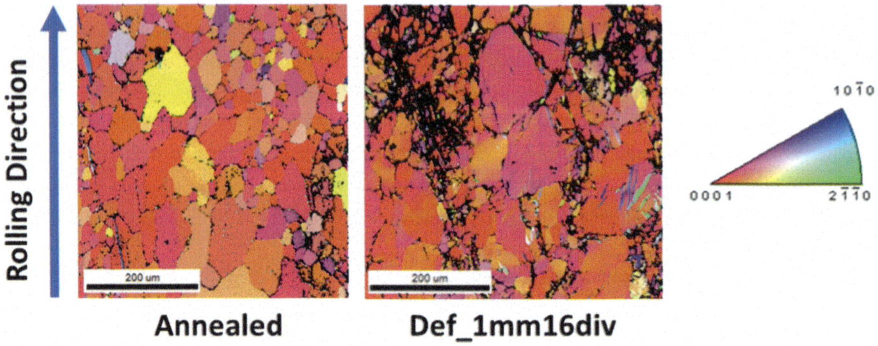

Fig. 14.2 IPF maps of sample before/annealed and after shock deformation. The colour-coded triangle on the right gives orientation of crystal in IPF map

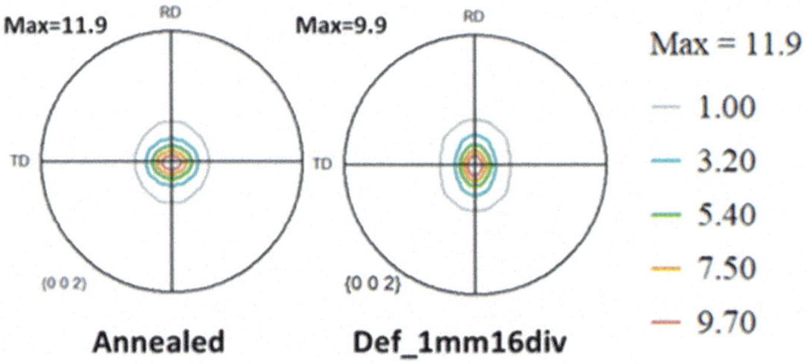

Fig. 14.3 (0001) pole figures for **a** annealed/undeformed and **b** deformed sample. The scale bar on right gives the intensity value of contours

Using the obtained EBSD data, micro strains of the shocked specimens are plotted as Kernal average misorientation (KAM) maps and plot, as shown in Fig. 14.4. Strain gradient within the grain having higher value at the grain boundary whereas peak shift in the kernel average misorientation (KAM) distribution plot by 0.5° (10%) towards higher strain in post deformed sample. It was observed that those grains which have more of a gradient in color show some strain within them. The strain was observed to be uniformly distributed within the grains. Also, it can be observed that there are some regions that were not indexed; this might be due to the higher amount of dislocation density there. In post shock cases, there are some grains that are completely strain-free; this might be because the twins who are formed in the grain must have grown, and the reorientation of the grain must have taken place.

Figure 14.5a shows that grain size distribution of sample before and after shock deformation. Shock induced deformation of pure polycrystalline Magnesium along perpendicular to basal plane resulted in grain refinement (Grain size decreases from 80 ± 18 μm to 45 ± 15 μm) mediated through formation of deformation twins (ET). Microhardness was calculated using the Vickers hardness test, which uses a square

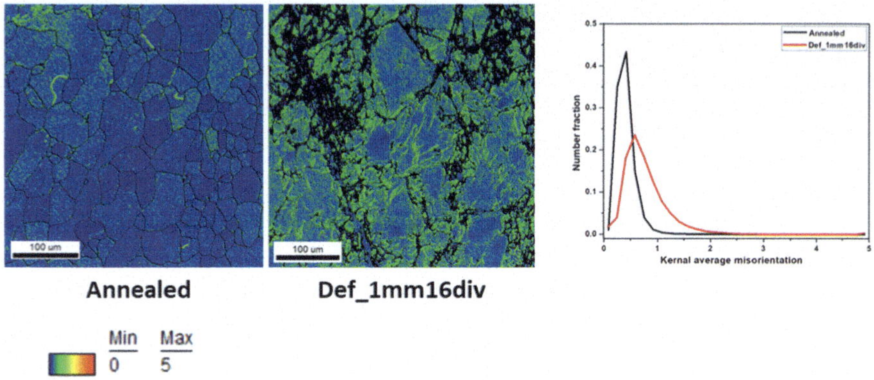

Fig. 14.4 Kernal average misorientation (KAM) maps and plot of sample before and after shock deformation

base diamond pyramid as an indenter. This test gives the hardness values in terms of Vickers Hardness Number (VHN). The shocked disc was cut along the diameter and mounted using a thermosetting powder. Hardness was taken at different lengths along the diameter, as shown in Fig. 14.5b. It can be observed that the hardness varied from one end of the diameter to another end of the diameter, as shown in Fig. 14.5b. Also, maximum hardness was seen where there is a maximum deflection in the shocked disc. The highest hardness is seen at the maximum deflection because where maximum deflection happens, the deformation there will be more, and more dislocation density will be present [13].

Figure 14.6 indicates that shock-deformed sample has shown similar misorientation distribution profile except for misorientation at angle 86°. Misorientation profile

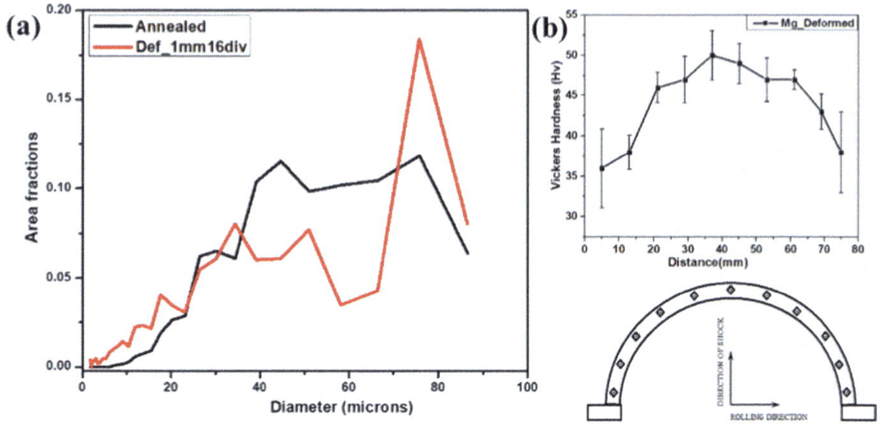

Fig. 14.5 **a** Grain size distribution of sample before and after shock deformation, **b** Vickers Hardness profile along RD direction through center on TD plane of shocked 80 cm circular disc

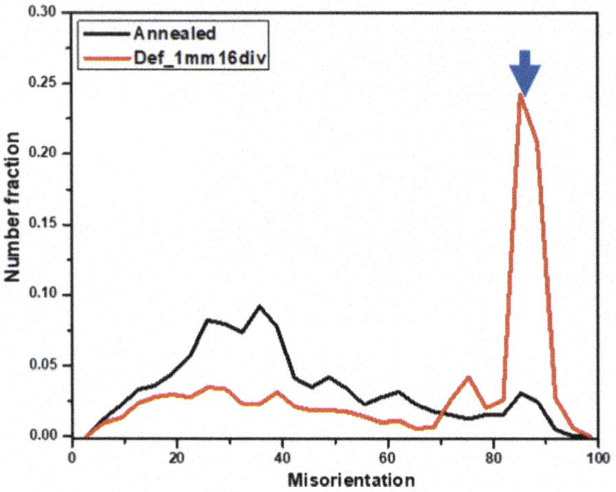

Fig. 14.6 Misorientation plot of annealed and deformed sample of pure magnesium

exhibits peak corresponding to twin misorientation. Initial sample have no peak at 86° which means there are no extensive twin in this sample. In deformed sample, peak is showing at 86° which means extensive twins are formed during deformation.

14.4 Conclusions

In the present study, the received pure magnesium block was unidirectionally rolled at elevated temperature and then annealed at 350 °C for 1 h. The prepared samples were shock-loaded, and microstructural and texture changes were observed. Upon analyzing the results done above, the following conclusion was drawn:

(a) It was observed that the strain at maximum deflection point was 6.52% at 0.8 Mpa peak pressure. This might be due to the activation of both twins and slips.
(b) Shock induced deformation of pure polycrystalline magnesium along perpendicular to basal plane resulted in grain refinement (80–40 μm) mediated through formation of deformation twins.
(c) There is a significant weakening of bulk texture has been observed because of grain fragmentation which produces many new orientations.
(d) Extension twinning (ET) of the $\{10\,\bar{1}\,2\}$ type at 86° misorientation angle and $<11\,\bar{2}\,0>$ misorientation axis are dominant deformation mechanism in shocked magnesium, also play key role in texture weakening. Mainly, basal (0001) texture weakening was seen, and no other components were observed.
(e) Strain gradient within the grain having higher value at the grain boundary whereas peak shift in the kernel average misorientation (KAM) distribution plot by 0.5° (10%) towards higher strain in post deformed sample.

(f) Microhardness was found to be highest (50 Hv) at the mid-point deflection region and it decreases systematically towards the periphery.

References

1. M.H. Yoo, S.R. Agnew, J.R. Morris, K.M. Ho, *Non-Basal Slip Systems in HCP Metals and Alloys: Source Mechanisms* (2001)
2. C. Bettles, M. Barnett, *Advances in Wrought Magnesium Alloys: Fundamentals of Processing, Properties and Applications*, vol. 1. Woodhead Publishing (2012)
3. T. Al-Samman, G. Gottstein, Dynamic recrystallization during high temperature deformation of magnesium. Mater. Sci. Eng., A **490**(1–2), 411–420 (2008)
4. G.I. Kanel et al., Shock response of magnesium single crystals at normal and elevated temperatures. J. Appl. Phys. **116**(14) (2014)
5. J.C.F. Millett, S.M. Stirk, N.K. Bourne, G.T. Gray, On the behaviour of the magnesium alloy, AZ61 to one-dimensional shock loading. Acta Mater. **58**(17), 5675–5682 (2010)
6. P. Renganathan, Y.M. Gupta, Shock compression/release of magnesium single crystals along a low-symmetry orientation: Role of basal slip. J. Appl. Phys. **126**(11) (2019)
7. H. Asgari, A.G. Odeshi, J.A. Szpunar, On dynamic deformation behavior of WE43 magnesium alloy sheet under shock loading conditions. Mater. Des. **63**, 552–564 (2014)
8. H. Asgari, A.G. Odeshi, J.A. Szpunar, L.J. Zeng, E. Olsson, Grain size dependence of dynamic mechanical behavior of AZ31B magnesium alloy sheet under compressive shock loading. Mater. Charact. **106**, 359–367 (2015)
9. K. Satheesh, G. Jagadeesh, K.P.J. Reddy, High speed schlieren facility for visualization of flow fields in hypersonic shock tunnels. Curr. Sci. **92**, 56–60 (2007)
10. A. Bisht, N. Ray, G. Jagadeesh, S. Suwas, Microstructural and crystallographic response of shock-loaded pure copper. J. Mater. Res. **32**(8), 1484–1498 (2017)
11. N. Ray, G. Jagadeesh, S. Suwas, Response of shock wave deformation in AA5086 aluminum alloy. Mater. Sci. Eng. A **622**, 219–227 (2015)
12. B. Beausir, S. Suwas, L.S. Tóth, K.W. Neale, J.J. Fundenberger, Analysis of texture evolution in magnesium during equal channel angular extrusion. Acta Materialia **56**(2), 200–214 (2008)
13. T. Inoue, H. Somekawa, T. Mukai, Hardness variation and strain distribution in magnesium alloy AZ31 processed by multi-pass caliber rolling. Adv. Eng. Mater. **11**(8), 654–658 (2009)

Chapter 15
Planar Shock Wave Moving Over a Double Concave Ramp

Sai Sharan Khandavilli, Karthik Chilkuri Reddy, and I. V. Thara Reshma

Abstract The objective of the present study is to understand the behavior of a normal shock wave, generated using a shock tube, traveling on a double concave ramp with a radius of curvature of 50 mm for different shock wave Mach numbers. An inviscid numerical simulation was carried out using Ansys Fluent to investigate the characteristics of the shockwave and the nature of the flow behind it. The unsteady shock behavior and flow characteristics were found to depend on the geometry and the shock wave Mach number.

15.1 Introduction

To understand the shock wave behavior and flow features, extensive analytical, experimental, and computational research on shock wave reflections and transitions in the steady, pseudo-steady, and unsteady frameworks have been undertaken throughout the years [1]. Studies on unsteady shock reflections were complex among the three, with the shock wave undergoing continuous transitions depending on the varying surface curvature and shock Mach number (M_s).

Most of the earlier studies in the unsteady flow regime concentrated on understanding the transition phenomena from regular (RR) to Mach reflection (MR) and vice versa, the RR $\leftrightarrow$ MR transitions, with some of the important works being the rigorous experimental studies by Takayama and Sasaki [2], analytical models by Ben-Dor and Takayama [3] and Hakkaki-Fard and Timofeev [4]. These studies were conducted on concave and convex cylindrical reflecting surfaces. RR $\rightarrow$ MR and MR $\rightarrow$ RR transition was experimentally investigated on coupled concave-convex and convex-concave surfaces by Geva et al. [5] and Ram et al. [6], respectively.

Considering similar geometry by Ram et al. [6], experimental and computational studies were carried out by Reshma et al. [7] on coupled convex-concave surface when a planar shock wave travels down the surface to understand the shock

S. S. Khandavilli · K. Chilkuri Reddy · I. V. Thara Reshma (✉)
Department of Aeronautical and Automobile Engineering, Manipal Institute of Technology, Manipal Academy of Higher Education, Manipal, Karnataka 576104, India
e-mail: thara.reshma@manipal.edu

© The Author(s), under exclusive license to Springer Nature Singapore Pte Ltd. 2024
G. Rajesh et al. (eds.), *Advances in Shock Interactions*, Springer Proceedings in Physics 408, https://doi.org/10.1007/978-981-97-6099-2_15

reflections and complex flow features. The shock reflections were found to be of a completely different nature when a planar shock wave moved down a ramp, with the formation of a stationary shock wave and vortex structures in the flow field, resembling shock motion in a converging–diverging nozzle. The work by Soni et al. [8] studied the shock reflection phenomenon as planar shock wave travels over a double concave reflector, in an upward direction, for different geometric conditions and surface angles. They found interesting shock reflections such as double Mach reflection (DMR) and transitional regular reflection (TRR) over the wedges. Considering the works by Reshma et al. [7] and Soni et al. [8], a numerical analysis was carried out in the present study to understand the shock reflection phenomena and flow behavior when a planar shock wave moves down a double concave geometry of equal radii, followed by a plane region.

15.2 Numerical Method

The computation is carried out using commercial CFD software—Ansys Fluent. An inviscid flow model was considered for the simulation. An implicit method has been presented for solving the singular initial value problem. The advection upstream splitting method (AUSM) flux calculation method was used as it provides exact resolution of contact and shock discontinuities. To improve the spatial accuracy for all types of meshes by reducing the numerical diffusion, third order MUSCL scheme was chosen for the reconstruction. The geometry was created as shown in Fig. 15.1, which depicts the downward double concave surface radius of 50 mm attached to a shock tube. The driver section of the shock tube is 300 mm while the driven section is 700 mm.

The shock tube boundaries are considered as walls and driver and driven sections were initialized with required pressure for required shock Mach numbers with air as the medium. A constant temperature of 300 K was taken in both sections.

The computational domain was divided into two parts with different grid sizes; a coarse mesh with an element size of 3 mm was used in the constant area section of the shock tube, and a more refined mesh with an element size of 0.5 mm was used over the ramp surface to refine the flow in the region. The computational domain

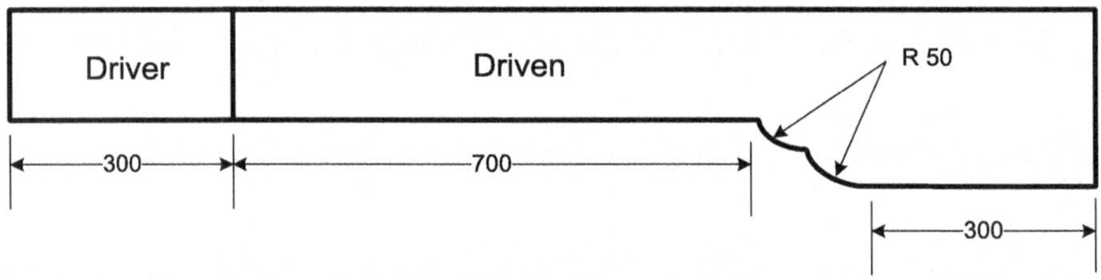

Fig. 15.1 Geometry of the computational domain (all dimensions in mm)

was discretized with a structured grid. For the meshing, the dominant quadrilateral setting was used along with body sizing and face sizing. The domain over the ramp surface was selected as the body of influence to get the maximum concentration of the grid over it. A grid independence study was carried out, and a grid of 414,304 elements and 415,524 nodes was chosen for the study.

15.3 Results

The objective of this study is to understand the behavior of a normal shock wave traveling over a double concave ramp of radius of curvature 50 mm at different shock Mach numbers. The nature of the shock wave and the flow features were examined in the study.

15.3.1 Shock Wave Reflections Over the Surface

The instantaneous behavior of the planar shock wave moving down the double concave surface at $M_s = 1.9$ is shown as a sequence of numerical schlieren images in Fig. 15.2. At time $t = 0$, the shock is located at the inlet of the first concave surface, as illustrated in Fig. 15.2a.

The starting point of the concave surface act as a sharp expansion corner with significant expansion waves formed, followed by a region of flow separation. As the shock wave is moving along the first concave surface, it bends in a concave nature with respect to the upstream flow to remain normal to the surface (see Fig. 15.2b). The curved shock wave positioned at the sharp intersection of the two concave surfaces generates a reflected shock wave forming an irregular reflection (IR) (see Fig. 15.2c). The IR wave, with a weak reflected shock wave, grows with time, with its curved Mach stem remaining perpendicular to the surface.

A distinct change in curvature is visible at the triple point (T_1) of the IR wave, as shown in Fig. 15.2d. As the shock wave moves further downstream, a secondary shock wave is formed on the first concave surface. The reflected shock wave of the IR also interacts with the separated flow region on the first concave surface (Fig. 15.2e). As the shock wave touches the plane region of the test section, another IR is formed at the foot of the curved Mach stem with a distinct triple point, T_2.

It is seen that the first reflected shock wave interacts with the flow in the first concave surface and the reflected shock formed at the planar surface interacts with that of the second concave surface (Fig. 15.2f). As the shock wave moves further downstream in the plane surface, there exist flow separation and vortex regions on both the concave surfaces. The reflected shock wave forms multiple interactions with the flow. A large region of flow separation is the reason for such shock interactions (Fig. 15.2g and h).

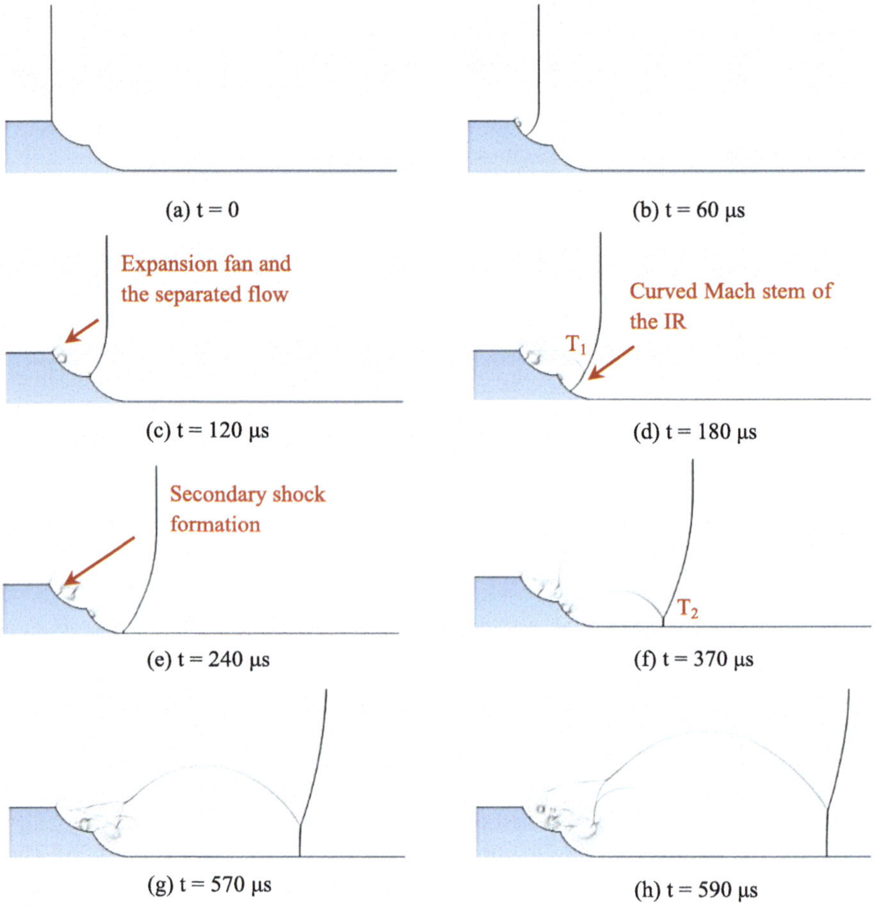

Fig. 15.2 Instantaneous numerical schlieren images of shock wave moving over a double concave surface at $M_S = 1.9$

15.3.2 Flow Features and the Stationary Shock Wave

The flow over the double concave surface showed significant features, such as the formation of a secondary shock wave and separation induced by the geometry. For a detailed study, the shock locations at times t = 330, 460, and 760 μs are considered where the planar shock wave is in the plane region of the test section. The velocity profile, surface pressure distribution, numerical schlieren image, and streamline distribution are shown in Figs. 15.3, 15.4 and 15.5, respectively.

The velocity profile and streamline distribution over the concave surfaces showed prominent flow separation regions (see Fig. 15.3a and 3d). A secondary shock wave formed at the first concave surface, and the interaction of the reflected shock wave with the flow is seen in Fig. 15.3c. The surface pressure plot in Fig. 15.3b shows

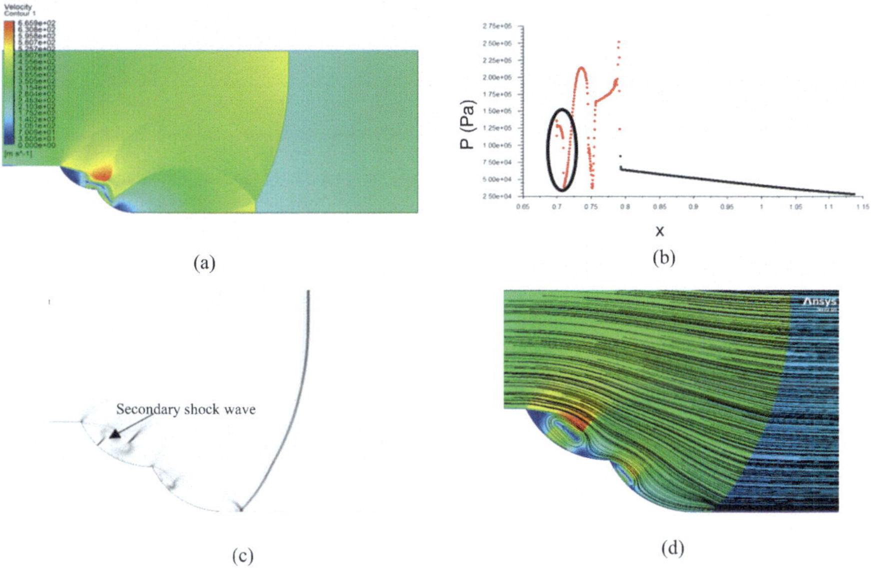

Fig. 15.3 Planar shock wave at t = 330 μs for M_s = 1.9 on the ramp surface. **a** The velocity profile, **b** surface pressure distribution, **c** numerical schlieren image with formation of stationary shock wave and **d** streamline distribution

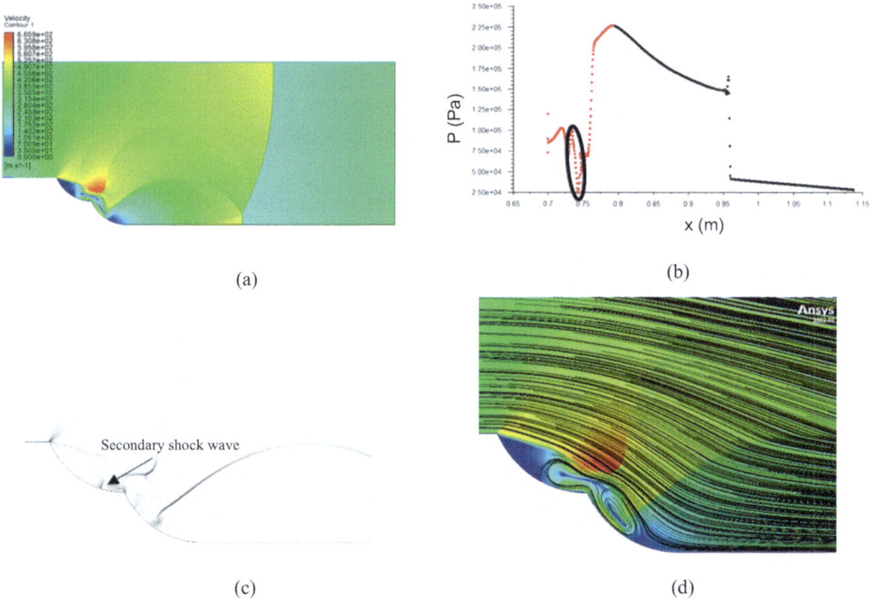

Fig. 15.4 Planar shock wave at t = 460 μs for M_s = 1.9 on the ramp surface. **a** The velocity profile, **b** surface pressure distribution, **c** numerical schlieren image with formation of stationary shock wave and **d** streamline distribution

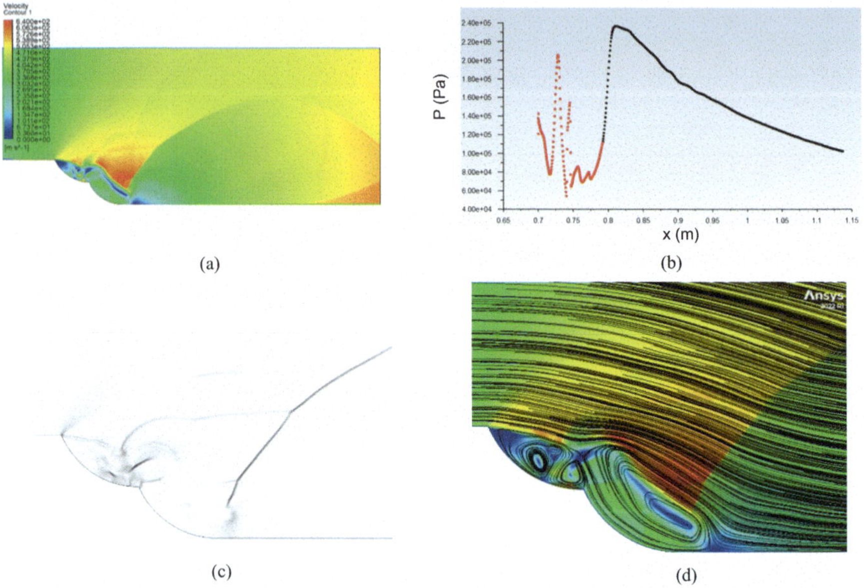

Fig. 15.5 Planar shock wave at t = 760 μs for M_s = 1.9 on the ramp surface. **a** The velocity profile, **b** surface pressure distribution, **c** numerical schlieren image with the formation of stationary shock wave, and **d** streamline distribution

a jump in pressure at the shock location with a magnitude of 3.25, whereas the primary shockwave had a pressure ratio of 4.06. Figure 15.3d shows the behavior of streamlines on the velocity contour. A large vortex was observed in the first cavity and a smaller size in the second cavity.

When the primary shockwave moved further to a position of 0.96 m from the diaphragm location, at t = 460 μs from the start of the double concave ramp, it was observed that the secondary shockwave also travelled further downstream. The velocity profile and numerical schlieren images are shown in Fig. 15.4a and c, respectively.

To understand the nature of the secondary shockwave, its shock strength was calculated from the pressure plot (Fig. 15.4b) and found to be 3.73, which is more than the initial pressure jump of 3.25 at t = 330 μs, Fig. 15.3b. The secondary shock wave moving downstream is also found to increase in strength, which also influences the flow on the surface. The vortices formed are also enlarged in both the concave surfaces with time (see Fig. 15.4d).

Another case was considered at time t = 760 μs when the main shock wave past the test section of the shock tube. Now, the secondary shock wave is no more present in the flow, whereas multiple shock interactions between the reflected shock wave and the separation bubble are visible, with a distinct 'Lamda' structure visible in the numerical schlieren image in Fig. 15.5c.

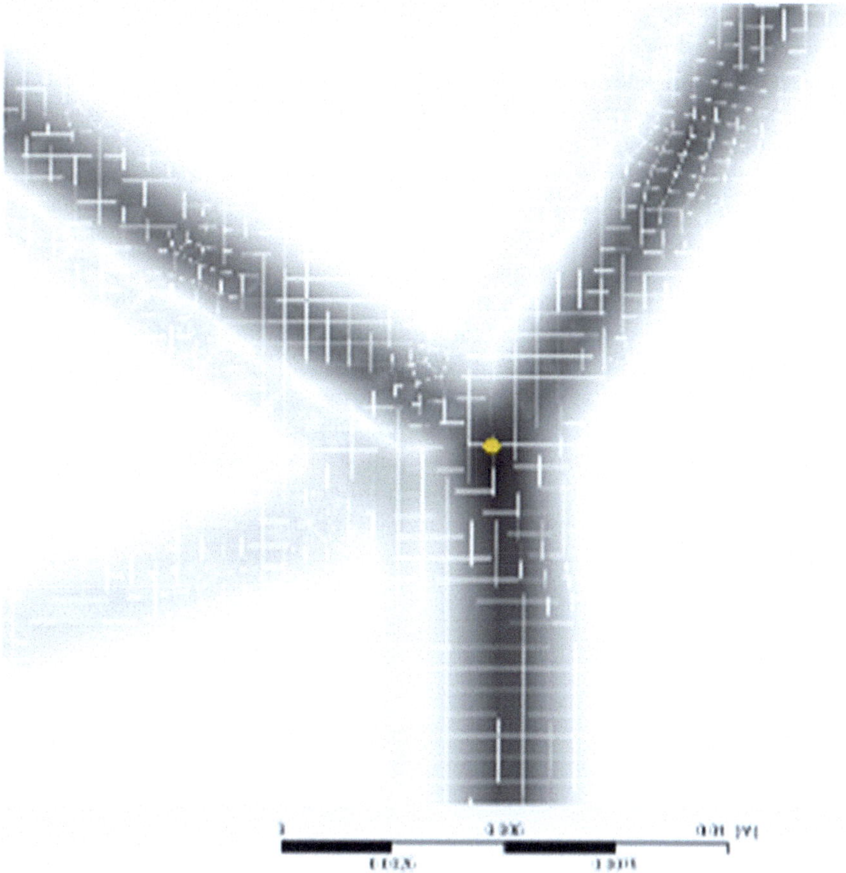

Fig. 15.6 The location of the triple point of the IR for $M_s = 1.4$

Velocity profile and streamline distribution signify multiple vortices formed in the first concave surface, followed by a large recirculation bubble filling the entire second concave surface in Fig. 15.5a and d, respectively. As the entire simulation was inviscid in nature, the flow separation observed was due to the sharp nature of geometry, at the beginning and between the two concave surfaces.

15.3.3 Triple Point Trajectory of the Irregular Reflection

As the planar shock wave moves past the double concave surface, in the plane region of the test section, it transforms into an IR with a prominent reflected shock wave and slip line. The IR wave grew with time as it moved downstream in the planar region (Fig. 15.2). The nature of the moving IR wave was studied primarily by tracing the trajectory of the triple point. In a pseudo-steady flow, the shock reflections are characterized by a linear change in the Mach Stem height, whereas in an unsteady flow, it is a nonlinear change [9].

The triple point locations at different instances were tracked from the numerical simulation results. An example of such a case is shown in Fig. 15.6, where the numerical schlieren image shows an instantaneous location of the shock with the triple point marked in yellow and the surface mesh visible in the background. A probe was used to get the approximate location of the triple point. As the size of each grid is 0.5 × 0.5 mm, the error in measurement is taken to be ± 0.5 mm.

The triple point trajectory of the IR wave for $M_s = 1.4$ is shown in Fig. 15.7, where the triple points are plotted as blue 'dots' for a time interval, dt = 21.6 μs. A second-order polynomial curve was fitted over the data point (solid line), showing the nonlinear TP trajectory of the IR.

The trajectory of the triple point for shock Mach numbers 1.2, 1.4, 1.6, 1.9, and 2.5 are plotted in Fig. 15.8, depicting the nonlinear TP trajectory along the straight portion of the test section. It is observed that the trajectory is curving up with a positive slope, indicating a continuously varying shock reflection or unsteady shock reflection on the test section [8]. The slope of the trajectory increased with increasing shock Mach number.

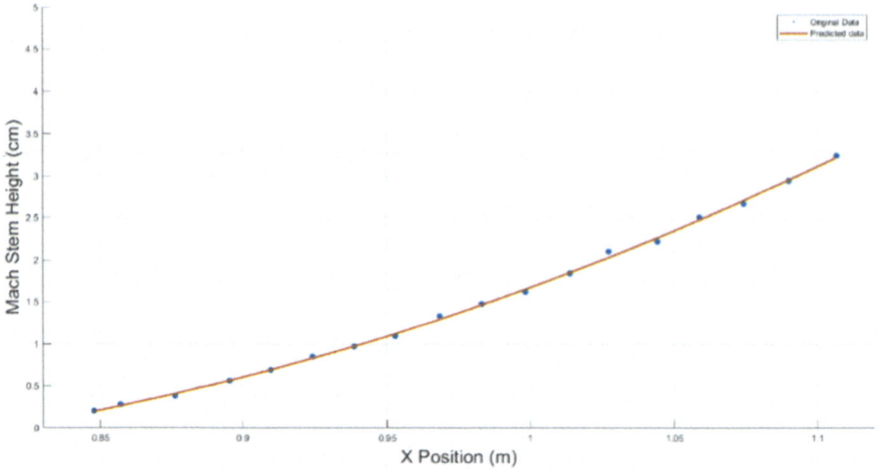

Fig. 15.7 Triple point trajectory of the IR with a second order polynomial curve fit plot for $M_s =$ 1.4

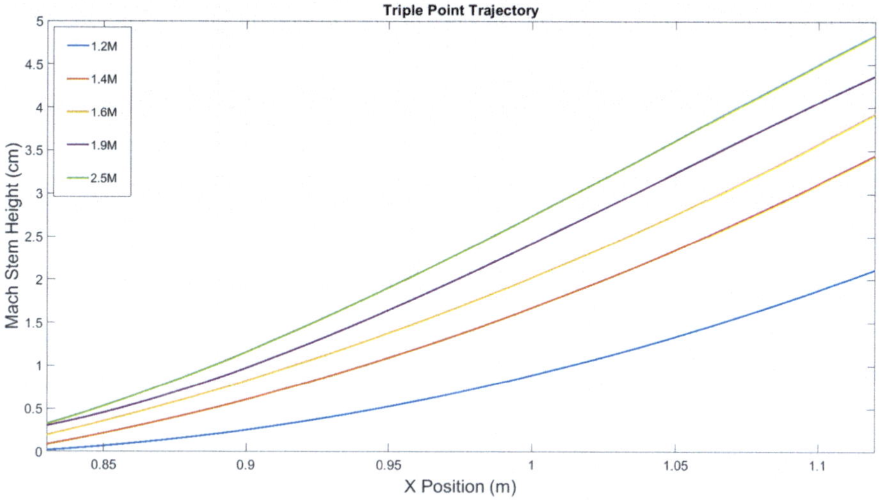

Fig. 15.8 Triple point trajectory of the IR wave at different M_S values

15.4 Conclusion

The present study investigates the reflections and transitions of a planar shock wave moving over a double concave surface of equal radii of 50 mm, followed by a plane section. The inviscid numerical analysis was carried out for shock Mach numbers of 1.2, 1.4, 1.6, 1.9, 2.3, and 2.5, using Ansys Fluent.

As the shock wave travels along the concave surface, it curves to remain normal to the first concave surface. On the second concave surface, the shock curvature has changed further with an irregular reflection pattern developed at the foot of the curved shock wave, with a curved Mach stem, remaining normal to the surface and a weak reflected shock wave. The shock wave now moves to the plane region of the test section, where a very prominent irregular reflection was formed with a distinct reflected shock wave and a slip line. The irregular reflection was observed to grow in an unsteady nature.

The flow analysis using streamlines and velocity contours predicted flow separation in both the concave surfaces, which grew with time. There was formation of secondary shock waves in the concave surfaces as the main shock wave past them, which interacted with the separated flow forming multiple reflection patterns were observed in the study. The unsteady shock movement in the concave surfaces caused the unsteady behavior of the irregular reflection in the plane region.

References

1. G. Ben-Dor, *Shock Wave Reflection Phenomena* (2007)
2. K. Takayama, M. Sasaki, Effects of radius of curvature and initial angle on the shock transition over concave or convex walls. Rept. Inst. High Speed Mech. Tohoku Univ. **46**, 1–30 (1983)
3. G. Ben-Dor, K. Takayama, Analytical prediction of the transition from Mach to regular reflection over cylindrical concave wedges. J. Fluid Mech. **158**, 365–380 (1985)
4. A. Hakkaki-Fard, E. Timofeev, On numerical techniques for determination of the sonic point in unsteady inviscid shock reflections. Int. J. Aerosp. Innov. **4**, 41–52 (2012)
5. M. Geva, O. Ram, O. Sadot, The non-stationary hysteresis phenomenon in shock wave reflections. J. Fluid Mech. **732**, R1 (2013)
6. O. Ram, M. Geva, O. Sadot, High spatial and temporal resolution study of shock wave reflection over a coupled convex-concave cylindrical surface. J. Fluid Mech. **768**, 219–239 (2015)
7. I. V. Thara Reshma, P. Vinoth, G. Rajesh, G. Ben-Dor, Propagation of a planar shock wave along a convex-concave ramp. J. Fluid Mech. **924**, 1–20 (2021). https://doi.org/10.1017/jfm.2021.631
8. V. Soni, A. Hadjadj, A. Chaudhuri, G. Ben-Dor, Shock-wave reflections over double-concave cylindrical reflectors. J. Fluid Mech. **813**, 70–84 (2017). https://doi.org/10.1017/jfm.2016.825
9. X. Yuan, J. Zhou, X. Mi, H.D. Ng, Model for triple-point trajectory of shock reflection over cylindrical concave wedge. AIAA J. **58**(6), 2770–2775 (2020). https://doi.org/10.2514/1.J059030

Chapter 16
Dynamics of Unsteady Shock Wave Reflections Using Overset Mesh Approach

Nikhil Jacob, P. J. Manoj, C. V. Sumesh Babu, A. J. Leo, and V. Lijo

Abstract The reflection of an oblique shock wave over a straight reflecting surface was investigated numerically in the classical inviscid gas-dynamics framework. A wedge in a steady supersonic flow was used to generate an oblique shock wave that was reflected from a symmetry plane. It is well known that for the steady reflection there exist only two possible configurations: the Regular Reflection (RR) and the Mach reflection (MR). The criteria of transition between them are well established for steady flows. In the present work, the leading edge of the wedge was subjected to slow/rapid rotation or periodic oscillation with different amplitudes creating unsteadiness in the flow. Such unsteadiness may cause premature RR $\rightarrow$ MR transition. However, little research has been conducted to show the effect of a continuous excitation on the dynamic development of shock wave reflections. For the rotation and oscillation of the wedge an overset framework is employed, where components are meshed individually (multiple cell zones) and typically embedded in a background mesh. The resulting multiple overlapping cell zones are connected by an overset interface. Overset grids maintain grid quality during mesh motion. The results for a flow Mach number $M = 4.0$ contradict the state-of-the-art knowledge regarding the RR $\rightarrow$ MR transition in steady flows, particularly for higher rates of wedge rotation. Well-conducted experimental data is often obtained in low-noise wind tunnels and numerical studies ensured disturbance-free computations to match with the experimental data. Unlike previous reports, for unsteady flows created by periodic perturbations, flow sustained a stable MR pattern well beyond its steady-state dual solution boundary.

N. Jacob (✉) · P. J. Manoj · C. V. Sumesh Babu · A. J. Leo · V. Lijo
Department of Mechanical Engineering, Government Engineering College Thrissur, Thrissur, Kerala 680009, India
e-mail: nikhilrit48@gmail.com

APJ Abdul Kalam Technological University, Thiruvananthapuram, Kerala, India

© The Author(s), under exclusive license to Springer Nature Singapore Pte Ltd. 2024

G. Rajesh et al. (eds.), *Advances in Shock Interactions*, Springer Proceedings in Physics 408, https://doi.org/10.1007/978-981-97-6099-2_16

16.1 Introduction

Shock reflection occurs when a steady supersonic flow encounters a wedge, which frequently appear in the intake of supersonic engines. If the reflecting wedge is positioned over a straight surface, the oblique shock wave will be reflected from the surface resulting in either a regular reflection (RR) or a Mach reflection (MR) wave configuration. Due to its influence on the flow, shock reflection has been intensively studied by Ben-dor in his past studies. Ernst Mach in 1878 has initially discovered the phenomenon of MR. Later Von Neumann in 1943 has carried out extensive research of MR in his researches. He observed that MR is made up of four branches that includes a slip stream and three shocks namely an incident shock I, a reflected shock R and a Mach stem. Two conditions proposed by various investigators for the RR $\rightarrow$ MR transition are the detachment condition beyond which an RR wave configuration is theoretically impossible and the von Neumann condition beyond which an MR wave configuration is theoretically impossible. Von Neumann was the first to propose these two conditions as possible RR $\rightarrow$ MR transition criteria.

Hornung and Robinson [1] showed that the RR $\rightarrow$ MR transition in steady flows depends on whether the incident Mach number M is greater than or equal to M_c, where M_c is the value at which transition lines of Von Neumann and detachment criteria intersect. It is calculated that the value of M_c for a perfect gas is 2.202. The intermediate domain, which is bounded by the von Neumann condition and the detachment condition, is known as the dual solution-domain. The solution domain in the Mach number -shock angle plane can be classified into three.

- A domain inside which only RR-configurations are theoretically possible.
- A domain inside which only MR-configurations are theoretically possible.
- A domain inside which both RR and MR-configurations are theoretically possible.

Hornung proposed that the existence of the dual solution domain implied there might be a hysteresis phenomenon at the transition. He assumed the transition from RR to MR would happen until the incidence angle increased to α_D. The Mach stem would arise suddenly and grow up at this angle. When the angle is decreased, the MR should persist throughout the dual solution domain and become RR at the angle α_N.

Markelov et al. [2] and Khotyanovsky et al. [3] investigated rapid wedge rotation with particular focus on the steady case. Investigations were done about the dynamics of the rapidly rotating wedge where the wedge was rotated about its leading edge at various rotational speeds for increasing and decreasing incident angles at $M = 3$.They also observed dynamic MR $\rightarrow$ RR transition at $M = 3.0$ briefly. Rapid wedge rotation resulted in MR $\rightarrow$ RR transition below the Von Neumann condition. So far there has been little research to consider the dynamic effect of a rapidly rotating wedge on the transition between regular and Mach reflection. They predicted significant deviation from the steady behaviour and concluded that the transition from Regular reflection to Mach reflection was easier to perform than the reverse transition.

Studies regarding the effect of rotational speed on the dynamic development of shock waves are scarce. In the present work, the effect of rotational speed of leading edge of the wedge in steady flow is numerically investigated. Numerical simulations were employed to study the transition effects from Regular to Mach Reflections. Hence it becomes necessary to investigate the effect of rotational speeds on shock wave reflections and to study the corresponding effects.

16.2 Computational Details

The basic dimensions of the wedge (Fig. 16.1) are taken from a study on two-dimensional wedge by Ivanov et al. [4]. The supersonic flow at inlet has a Mach number of 4.0, temperature 300 K. The overall pressure is assumed as 1.0 bar. The working fluid is assumed as dry air. The reflection of an oblique shock from a wedge on a perfect reflecting plane is discussed here, without the complicating feature of a boundary layer. Hence, an Euler CFD code is used to simulate the rapidly rotating wedge.

The mesh consists of 0.3 million purely structured quad elements, and the time-step selected is 4×10^{-7} s after grid and time-step independence studies. Details of the computational mesh are shown in Fig. 16.2. For the rotation and oscillation of the wedge an overset framework is employed, where wedge is meshed individually (foreground mesh) and typically embedded in a background mesh. The resulting multiple overlapping cell zones are connected by an overset interface. Overset grids maintain grid quality during mesh motion, enabling us to give high rate of rotation than possible with dynamic/deforming meshes.

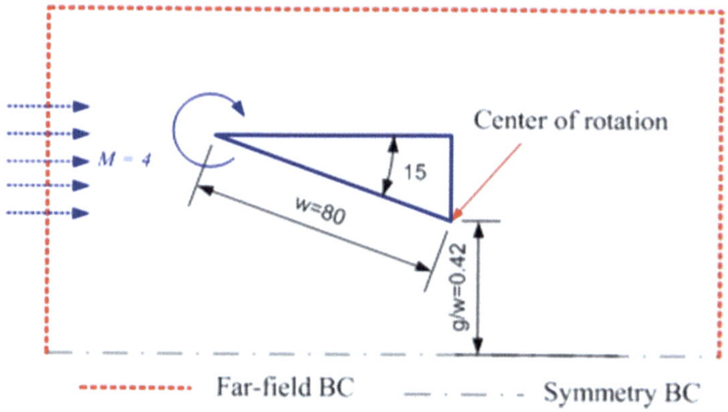

Fig. 16.1 Computational domain with boundary conditions

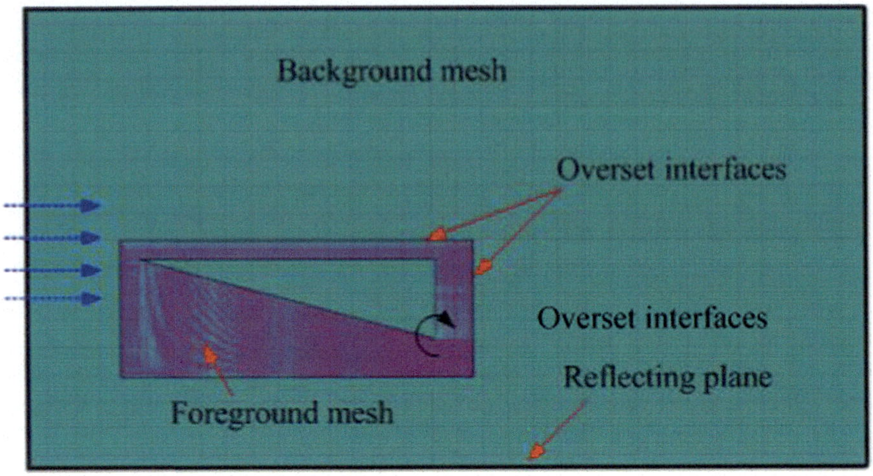

Fig. 16.2 Computational mesh

16.3 Results and Discussion

Comparison of the numerical results of the wedge simulation with those obtained from the oblique shock theory (Table 16.1) shows excellent agreement. The error in static pressure prediction is less than 0.5%, while error in shock angle prediction is about 5% [4].

The shock reflection from the rotating wedge investigated and detailed through instantaneous numerical schlieren in Fig. 3b, which demonstrates the proper behavior when compared to the experimental schlieren [4]. Hence the present overset grid formulation is sufficient for predicting the complex system of shock and expansion waves during rotation of the wedge.

Figure 16.4 shows the transient development of Mach stem with rotational speeds. For the same wedge angle (Θ_w), slower rotational speed shows greater Mach stem height. Moreover, at higher rotational speeds, the shock transition to MR occurs at larger wedge angle.

Table 16.1 Comparison of the numerical results with oblique shock theory

	Mach number	Static pressure (bar)	Temperature (K)	Density (kg/m^3)	Shock angle (°)
Oblique-shock theory	2.209	7.054	637.63	3.855	38.46
Fluent solution	2.172	7.092	647.73	3.893	40.46
Percent difference	1.67	0.54	1.58	0.98	5.2

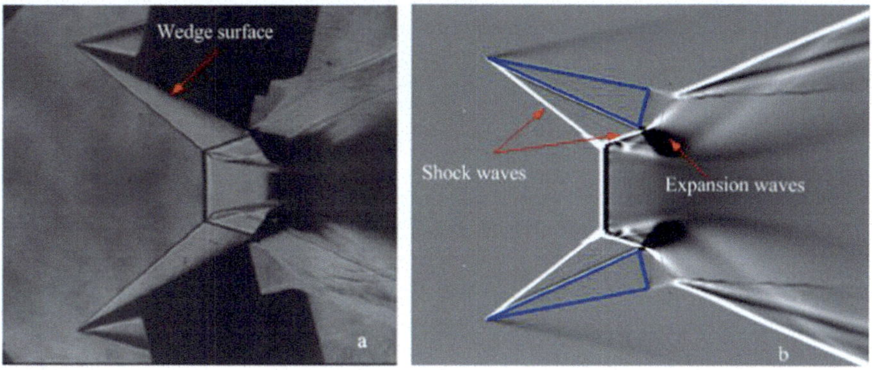

Fig. 16.3 **a** Experimental Schlieren [4], **b** Numerical Schlieren with 50°/s rotation of wedge for the same flow conditions

Wedge rotational speed greatly influences the Mach stem growth as shown in Fig. 16.5. For a very slow rotational rate of 5^0/s, the growth of Mach is gradual, whereas higher rotational speeds accelerate Mach stem growth. 200^0/s and 400^0/s show similar Mach disk growth trend. For the slow rotational rate of 5^0/s, the establishment of MR is also delayed. Present results show that the RR- > MR transition is greatly influenced by the rate of rotation of the wedge. Results from the unsteady simulations with wedge subjected to periodic oscillation with different amplitudes will be reported in future studies.

Fig. 16.4 Instantaneous Mach contours for various rotational speed of the wedge. Angle of the wedge is same (26°) for comparison

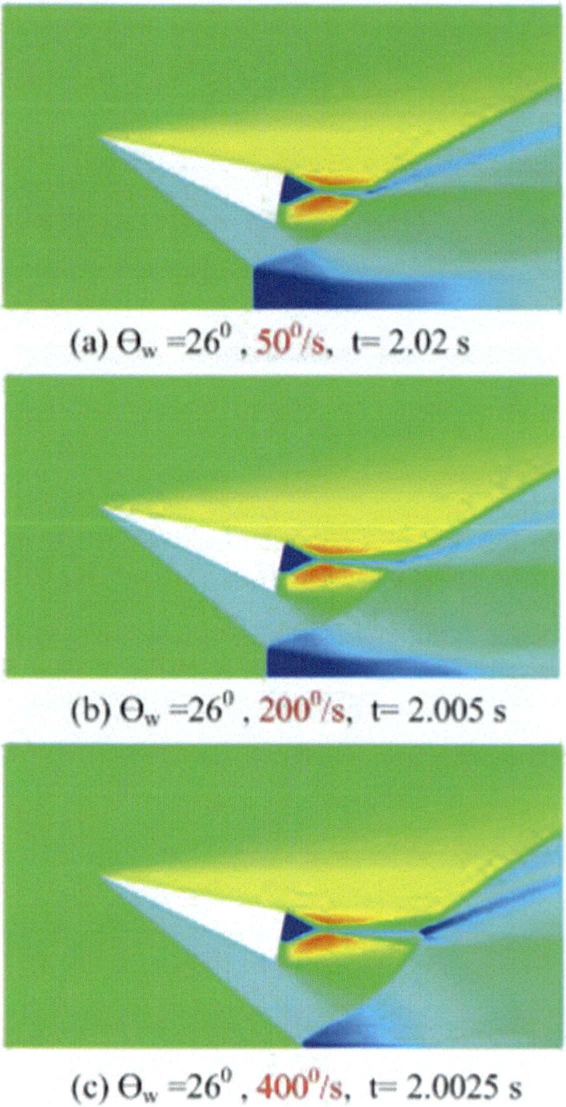

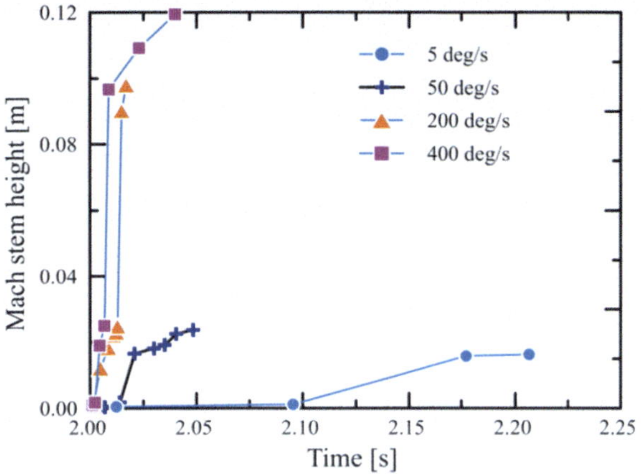

Fig. 16.5 Effect of wedge rotational rate on Mach stem growth

16.4 Conclusion

The present work numerically investigates the dynamic shock transition of shock waves incident on a perfect reflecting plane. The incident shock is generated by a wedge, which can be rotated. As the wedge angle changes, the incident shock angle also changes, bringing an RR → MR transition. An Euler CFD code is used for the simulations, as we don't want to contaminate the shock reflection point by considering viscous boundary layer. Overset mesh formulation is employed for the slow/rapid rotation of the wedge. Results obtained with overset mesh are validated against oblique shock theory and experimental schlieren, which shows good match. Present results show that the RR → MR transition is greatly influenced by the rate of rotation of the wedge. Wedge rotational speeds of 200^0/s and 400^0/s show similar Mach disk growth trend. Whereas, for slow rotational speed of 5^0/s, the establishment of MR is delayed.

References

1. Hornung HG , Robinson ML, Transition from regular to Mach Reflection of Shock Waves. J Fluid Mech **123**, 155–164 (1982)
2. Ivanov MS, Markelov GN, Kudryatsev AN, Numerical analysis of shock wave reflection transition in steady flows. AIAA J. 2079–2086 (1998)
3. Khotyanovsky DV ,Kudryavstev AN, Ivanov MS, Numerical study of transition between steady regular and Mach reflection caused by free stream pertubrations (1999)
4. M.S. Ivanov, D. Vandromme, V.M. Fomin, A.N. Kudryavtsev, A. Hadjadj, D.V. Khotyanovsky, Transition between regular and Mach reflection of shock waves, new numerical and experimental results. Shock Waves **407**, 85–104 (2001)

Chapter 17
Shock-Granular Bed Interaction and Effect of Compaction Models

Annie Rose Elizabeth and T. Jayachandran

Abstract Shock transiting through energetic granular bed is found during operations of gun propulsion systems, actuators, pin pullers, pyrotechnic igniters, etc. The operation time of these systems are very short and study of shock interaction with the granular bed is crucial. In this paper, the effects of shock on dense granular bed like particle compaction, shock reflection, shock transmission, etc. are studied. For this study, six equation model in compressible multifluid formulation is used with and without compaction model. This model is thermodynamically consistent and unconditionally hyperbolic. This model could accurately model the propagation of pressure, velocity and density disturbances. A two-dimensional axisymmetric code is developed using this model for the study. Unstructured finite volume method with second order accurate HLLC scheme is used for spatial discretization and classical four stage Runge–Kutta method as explicit method for temporal discretization.

17.1 Introduction

Shock transiting through energetic granular bed is found during operations of gun propulsion systems, actuators, pin pullers, pyrotechnic igniters, etc. The operation time of these systems are very short and study of shock interaction with the granular bed is crucial. In this paper, the effects of shock on dense granular bed like shock reflection, shock transmission, particle compaction, etc. are studied. To study the effects, Rogue test case [1] is utilized. In a long at there is a 2 cm thick granular bed glass particle. A shock is passed through a dense granular bed. The pressure is monitored ahead of the granular bed to study the incident and reflected shock. The pressure is monitored behind the granular bed is tracked to study the transmitted shock. As the shock transits the bed, the change in the particle volume fraction is analyzed to study the effects of compaction. The work presented focuses on the numerical study of shock interaction with a dense granular bed. The strength of the incident shock, reflected shock and transmitted shock is studied. Also, the effect

A. R. Elizabeth (✉) · T. Jayachandran
Indian Institute of Technology Madras, Chennai, Tamil Nadu, India
e-mail: annie29992@gmail.com

© The Author(s), under exclusive license to Springer Nature Singapore Pte Ltd. 2024

G. Rajesh et al. (eds.), *Advances in Shock Interactions*, Springer Proceedings in Physics 408, https://doi.org/10.1007/978-981-97-6099-2_17

of having a compaction model in the governing equation is studied. Interaction of shock waves to granular bed has been studied for a very long time. The effects of the interaction on dilute bed and dense bed are entirely different [2]. The dynamics of reflected shocks was studied by Skews [3]. The compaction of granular bed was studied by Kuo et al. [4], Powers et al. [5], Gonthier et al. [6], Bdzil et al. [7], Garcia-Cascales et al. [8], Zhen et al. [9], Saurel et al. [10], Gambino et al. [11], McGrath et al. [12], Rao [13], Lopez-Munoz et al. [14], etc. Various compaction models were compared and three models selected for the present study are [9, 10, 14].

For this study, six equation model in compressible multifluid formulation is used with and without compaction model. This model could accurately model the propagation of pressure, velocity and density disturbances. A two-dimensional axisymmetric code is developed using this model for the study. Unstructured finite volume method is used for spatial discretization and classical four stage Runge–Kutta method as explicit method for temporal discretization. For flux computation second order accurate HLLC scheme is used. The gas and particle phases are considered to be compressible through suitable equation of state. The subsequent sections set out the governing equations with the compaction models, numerical model, problem description, results with discussion and finally conclusion of the present study.

17.2 Mathematical Model

17.2.1 Governing Equation

The compressible multi-fluid formulation using six-equation model [15] in axisymmetric two dimension is as follows. In the current problem of study, there is both gas phase and particle phase. Both phases are considered to be compressible, evolving with its own velocity, temperature and pressure. The gas phase is governed by ideal gas equation of state and particle phase is governed by stiffened gas equation of state. In six-equation model, it is assumed that the gas phase (g) and particle phase (p) pressures are same ($P_p = P_g$), if there is no compaction. In the case of compaction, it is assumed that particle phase pressure is the sum of gas phase pressure and compaction pressure ($P_p = P_g + \beta$). This model takes care of the wave propagation in both particle as well as gaseous phase.

The governing equations are written in the generic form as where U is the conserved variables, F, G are the flux terms in x, y direction respectively and S is the source term. Suffix g, p, I denote gas phase, particle phase and interface respectively. $\emptyset, \rho, u, E, P, \beta$ denote volume fraction, density, velocity, total energy, pressure and compaction pressure respectively.

$$\frac{\partial U}{\partial t} + \frac{\partial F}{\partial x} + \frac{\partial G}{\partial y} = S \tag{17.1}$$

where,

$$U = \begin{bmatrix} \emptyset_g\rho_g \\ \emptyset_g\rho_g u_g \\ \emptyset_g\rho_g v_g \\ \emptyset_g\rho_g E_g \\ \emptyset_p\rho_p \\ \emptyset_p\rho_p u_p \\ \emptyset_p\rho_p v_p \\ \emptyset_p\rho_p E_p \end{bmatrix}; F = \begin{bmatrix} \emptyset_g\rho_g u_g \\ \emptyset_g\rho_g u_g^2 + \emptyset_g P_g \\ \emptyset_g\rho_g u_g v_g \\ \emptyset_g u_g\rho_g E_g + \emptyset_g u_g P_g \\ \emptyset_p\rho_p u_p \\ \emptyset_p\rho_p u_p^2 + \emptyset_p P_p \\ \emptyset_p\rho_p u_p v_p \\ \emptyset_p u_p\rho_p E_p + \emptyset_p u_p P_p \end{bmatrix}; G = \begin{bmatrix} \emptyset_g\rho_g v_g \\ \emptyset_g\rho_g u_g v_g \\ \emptyset_g\rho_g v_g^2 + \emptyset_g P_g \\ \emptyset_g v_g\rho_g E_g + \emptyset_g v_g P_g \\ \emptyset_p\rho_p v_p \\ \emptyset_p\rho_p u_p v_p \\ \emptyset_p\rho_p v_p^2 + \emptyset_p P_p \\ \emptyset_p v_p\rho_p E_p + \emptyset_p v_p P_p \end{bmatrix};$$

$$S = \begin{bmatrix} 0 \\ D_x + P_I\frac{\partial\emptyset_g}{\partial x} \\ D_y + P_I\frac{\partial\emptyset_g}{\partial y} + \frac{\emptyset_g(P_g - \sigma_{\theta\theta,g})}{y} \\ U_I D_x + V_I D_y + P_I U_I\frac{\partial\emptyset_g}{\partial x} + P_I V_I\frac{\partial\emptyset_g}{\partial y} \\ 0 \\ -D_x - P_I\frac{\partial\emptyset_g}{\partial x} \\ -D_y - P_I\frac{\partial\emptyset_g}{\partial y} + \frac{\emptyset_p(P_p - \sigma_{\theta\theta,p})}{y} \\ -U_I D_x - V_I D_y - P_I U_I\frac{\partial\emptyset_g}{\partial x} - P_I V_I\frac{\partial\emptyset_g}{\partial y} \end{bmatrix}$$

The volume fractions should satisfy

$$\emptyset_g + \emptyset_p = 1 \tag{17.2}$$

The gas phase is governed by ideas gas equation of state:

$$P_g = \rho_g R T_g \tag{17.3}$$

The particle phase is governed by stiffened gas equation of state [10]:

$$P_p = \rho_p(\gamma_p - 1)e_p - \gamma_p P_\infty \tag{17.4}$$

The common pressure P is found using the volume fraction relation (17.2) as mentioned by Yeom et al. [15]. The bounding equations are given by Elizabeth et al. [16]. Once the common pressure is computed, the gas and particle phase pressures are as follows and if compaction is not modelled then, β becomes zero.

$$P_g = P \tag{17.5}$$

and

$$P_p = P_g + \beta \tag{17.6}$$

The compaction models used for including compaction pressure (β) are given below.

Model 1 [14]:

$$\beta = \begin{cases} \frac{48.8\times 10^6}{(1-\varnothing_p)}\left(\frac{1}{(1-\varnothing_{p,c})} - \frac{1}{(1-\varnothing_p)}\right) & if\ \varnothing_p \le \varnothing_{p,c} \\ 0 & if\ \varnothing_p > \varnothing_{p,c} \end{cases} \tag{17.7}$$

where $\varnothing_{p,c} = 0.72$.

Model 2 [9]:

$$\beta = \begin{cases} \frac{4\times 6.7\times 10^7(\varnothing_{p,0}-\varnothing_p)}{3\varnothing_p(\varnothing_p-1)} & if\,\varnothing_{p,0} \ge \varnothing_p > \varnothing_{p,1} \\ \frac{2}{3}5.9\times 10^6\left[1 - \frac{2\times 6.7\times 10^7}{5.9\times 10^6\varnothing_p}(\varnothing_{p,0}-\varnothing_p) + \ln\left(\frac{2\times 6.7\times 10^7(\varnothing_{p,0}-\varnothing_p)}{5.9\times 10^6(\varnothing_p-1)}\right)\right] & if\,\varnothing_{p,1} \ge \varnothing_p > \varnothing_{p,2} \\ \frac{2}{3}5.9\times 10^6 \ln\frac{\varnothing_p}{\varnothing_p-1} & if\,\varnothing_{p,2} \ge \varnothing_p > 1 \end{cases} \tag{17.8}$$

where $\varnothing_{p,0}$ is the initial particle volume fraction,

$$\varnothing_{p,1} = \frac{2\times 6.7\times 10^7\varnothing_p + 5.9\times 10^6}{2\times 6.7\times 10^7 + 5.9\times 10^6} \tag{17.9}$$

$$\varnothing_{p,2} = \frac{2\times 6.7\times 10^7\varnothing_p}{2\times 6.7\times 10^7 + 5.9\times 10^6} \tag{17.10}$$

Model 3 [10]:

$$\beta = \begin{cases} -3\times 10^4\times 1.02\varnothing_p\rho_p(1+\log(1-\varnothing_p))(C_1+C_2-C_3)^{(1.02-1)/1.02} & if\,\varnothing_{p,0} < \varnothing_p < 1 \\ 0 & otherwise \end{cases} \tag{17.11}$$

where

$$C_1 = (1-\varnothing_p)\log(1-\varnothing_p) \tag{17.12}$$

$$C_2 = (1+\log(1-\varnothing_{p,0}))(\varnothing_p-\varnothing_{p,0}) \tag{17.13}$$

$$C_3 = (1-\varnothing_{p,0})\log(1-\varnothing_{p,0}) \tag{17.14}$$

17.3 Numerical Model

The mathematical model is basically partial differential equations (PDEs) with time and space dependency. To numerically solve it, first the PDEs are integrated in space to obtained a set of ordinary differential equations (ODEs). The ODEs are integrated in time to obtain the solution. Finite volume method is used for spatial discretization and explicit four stage Runge–Kutta method for temporal discretization. Second order accurate HLLC scheme is used for flux computation.

Equation (17.1) can be rewritten in the form

$$\frac{\partial U}{\partial t}+\frac{\partial F}{\partial x}+\frac{\partial G}{\partial y}=\frac{\partial \boldsymbol{U}}{\partial t}+\nabla.\boldsymbol{E}=S \tag{17.15}$$

Upon spatial integration, the PDE becomes

$$\frac{d\boldsymbol{U}_i}{dt}=-\frac{1}{\Omega_i}\left[\oint\!\!\!\oint \boldsymbol{E}_i dA_i-\textstyle\int S_i d\Omega_i\right]=-\frac{1}{\Omega_i}R_i \tag{17.16}$$

The HLLC scheme used here for two phase flows is derived from single phase HLLC scheme mentioned by Batten et al. [17]. It is considered to have two averaged intermediate states $\boldsymbol{U}_L^*$ and $\boldsymbol{U}_R^*$, separated by the contact wave with speed S_M. Figure 17.1 depicts the four averaged states with two intermediate states separated by contact wave. The acoustic waves with largest and smallest velocities are denoted as S_R and S_L respectively. If $S_L > 0$, then the flow is supersonic from left to right. If $S_R < 0$, then the flow is supersonic from right to left. The flow is subsonic in the region $S_L < 0 < S_R$.

Summarizing the HLLC scheme to compute the fluxes at the cell interface where L and R denotes left and right respectively.

$$\boldsymbol{E}_{HLLC}=\begin{cases}\boldsymbol{E}_L; & S_L>0\\ \boldsymbol{E}_L+S_L\left(\boldsymbol{U}_L^*-\boldsymbol{U}_L\right); & S_L\le 0<S_M\\ \boldsymbol{E}_R+S_R\left(\boldsymbol{U}_R^*-\boldsymbol{U}_R\right); & S_M\le 0\le S_R\\ \boldsymbol{E}_R; & S_R<0\end{cases} \tag{17.17}$$

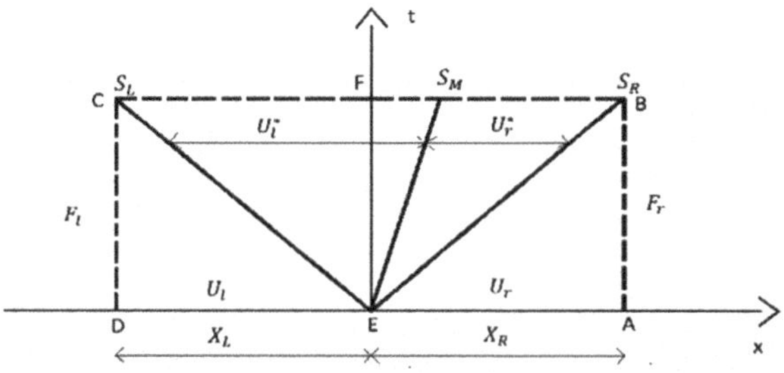

Fig. 17.1 Simplified Riemann fan with two intermediate states

where,

$$S_L = \min\left(q_{g,L} - a_{g,L}, q_{g,R} - a_{g,R}, q_{p,L} - a_{p,L}, q_{p,R} - a_{p,R}\right) \tag{17.18}$$

$$S_R = \max\left(q_{g,L} + a_{g,L}, q_{g,R} + a_{g,R}, q_{p,L} + a_{p,L}, q_{p,R} + a_{p,R}\right) \tag{17.19}$$

$$q_{g,k} = u_{g,k}n_x + v_{g,k}n_y \quad ; \quad q_{l,k} = u_{l,k}n_x + v_{l,k}n_y \quad (k = L,\ R) \tag{17.20}$$

$$S_M = \frac{\rho_R q_R (S_R - q_R) - \rho_L q_L (S_L - q_L) + P_L - P_R}{\rho_R (S_R - q_R) - \rho_L (S_L - q_L)} \tag{17.21}$$

$$P_g^* = P_p^* = P_L + \rho_L (q_L - S_L)(q_L - S_M) \tag{17.22}$$

$$P_L = \emptyset_{g,L} P_{g,L} + \emptyset_{p,L} P_{p,L} \tag{17.23}$$

$$\rho_L = \emptyset_{g,L} \rho_{g,L} + \emptyset_{p,L} \rho_{p,L} \tag{17.24}$$

$$q_L = \left(\emptyset_{g,L} \rho_{g,L} q_{g,L} + \emptyset_{p,L} \rho_{p,L} q_{p,L}\right) / \rho_L \tag{17.25}$$

Denoting $k = L, R$ and $m = g, p$ for the subsequent equations where,

$$\boldsymbol{U}^*_{k,m} = \emptyset_{k,m} \rho_{k,m} \frac{\left(S_{k,m} - q_{k,m}\right)}{\left(S_{k,m} - S_M\right)} \begin{bmatrix} 1 \\ u_{k,m} \\ v_{k,m} \\ E_{k,m} \end{bmatrix} + \frac{1}{\left(S_{k,m} - S_M\right)} \begin{bmatrix} 0 \\ \left(\emptyset^*_{k,m} P^*_m - \emptyset_{k,m} P_{m,k}\right) n_x \\ \left(\emptyset^*_{k,m} P^*_m - \emptyset_{k,m} P_{m,k}\right) n_y \\ \emptyset^*_{k,m} P^*_m S_M - \emptyset_{k,m} P_{m,k} q_{k,m} \end{bmatrix} \tag{17.26}$$

17.4 Problem Description

To carry out the present study, the developed code is validated against the Rogue test case [1]. A 3.6 m long, 13 cm diameter cross section shock tube with 0.8 m of driver section and 2.8 m of driven section is taken for the study. At 1.150 m there is a 2 cm thick granular bed of spherical glass particles. For the validation, a 1.3 Mach shock is passed through the granular bed of 0.65 particle volume fraction with 1.5 mm diameter glass particles. Pressure at 11 cm ahead of the granular bed is validated

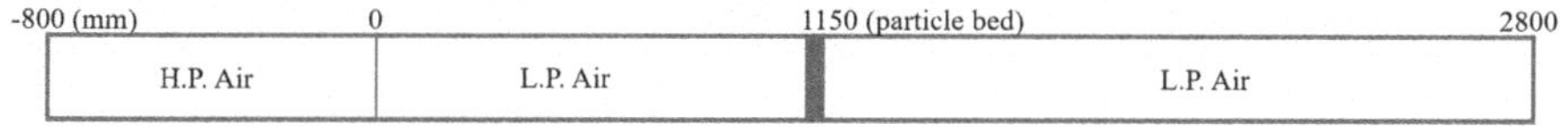

Fig. 17.2 Computational domain

Table 17.1 Study cases

Case No.	Incident shock Mach number	Particle volume fraction	Particle diameter (mm)
1	1.3	0.65	1.5
2	1.6	0.65	1.5
3	1.95	0.65	1.5
4	1.3	0.65	0.75
5	1.95	0.65	0.75
6	1.3	0.65	3.0
7	1.95	0.65	3.0
8	1.3	0.73	1.5
9	1.95	0.73	1.5

with Rogue et al. [1]. The validated code is then used to study the interaction of shock on a granular bed and the effects of compaction. The pressure is monitored 11 cm (P1) ahead of the granular bed to study the incident and reflected shock. The pressure at 71 cm (P2) after the granular bed is tracked to study the transmitted shock. The computational domain is exhibited in Fig. 17.2. As the shock transits the bed, the change in the particle volume fraction is analyzed to study the effects of compaction. Wall boundary condition is used on the three sides of the shock tube and symmetry boundary condition on the y-symmetry side of the shock tube. The initial pressure in the driven section is 100 kPa and the initial temperature of both the phases are 308.15 K. The density of the gas phase is 1 kg/m^3 and that of particle phase is 2500 kg/m^3. The specific heat ratio for the gas phase is 1.4 and that of particle phase is 2.5. The pressure constant P_∞ is 0.1 GPa. The various cases used for the study are summarized in Table 17.1. To generate a shock of Mach 1.3, the driver pressure is initialized to 345.18 kPa. Likewise for Mach 1.6 and 1.95 the driver pressures are 975.78 kPa and 2902.79 kPa respectively. The cases mentioned in Table 17.1 are studied with and without the compaction models.

17.5 Results and Discussion

First, the code is validated with Rogue et al. [1] and is shown in Fig. 17.3. In Fig. 17.3, the pressure history at 11 cm ahead of the granular bed is compared for the models with all the three compaction models and without compaction. After validation, the code has been used to study the shock interaction with the granular bed and the effects of compaction models. Pressure histories at P1 is studied for Case 1 and Case 3 (mentioned in Table 17.1) as shown in Figs. 17.3 and 17.4. The first rise in pressure is when the incident shock wave passes through P1. The second rise in pressure is when the refracted shock wave passes through P1. The shock strength has increased as the incident shock speed increases. From the three compaction models, it observed that Model 3 is better in capturing the physics. Henceforth for all the studies Model 3 is used. Figures 17.5 and 17.6 shows the pressure histories for Case 1, 2 and 3 at P1 and P2. It is observed that the shock strength gets higher as the Mach number increases. The incident shock and reflected shock is least affected by the compaction model. But the transmitted shock strength is slightly lesser when compaction is accounted. The same can be observed in Case 1, 4, 6 and 8. Even for higher Mach number, with different particle diameter and volume fraction, the shock interaction with the bed is studied and results in similar nature is observed. To study the effect of compaction model on the granular bed, particle volume fraction along the axial direction is plotted. Figures 17.7 and 17.8 shows the shock wave interaction with the granular bed for without and with compaction model respectively for Case 1 (Mach 1.3) at 0, 2, 2.2, 2.5 and 3 ms. The same for Case 2 (Mach 1.6) is shown in Figs. 17.9 and 17.10 at 0, 1.5, 1.8, 2.0 and 2.5 ms. When compaction is not accounted for, the granular bed particle volume fraction increases as the shock transits the bed. When compaction is accounted as shock transits the bed, a slight displacement of the granular bed is observed with a reduction in the bed thickness while the particle volume fraction in the granular bed remains the same as initialized.

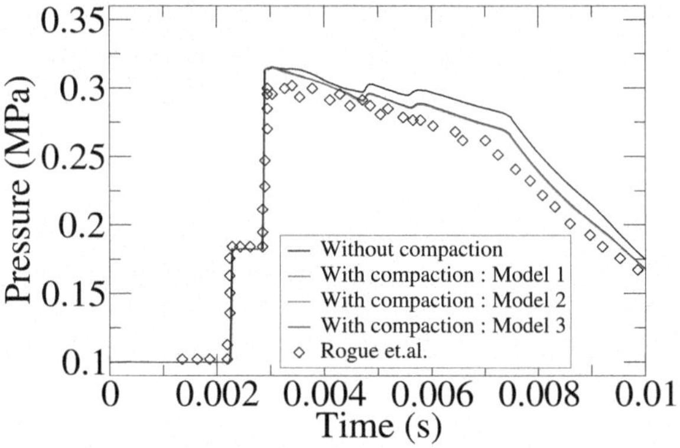

Fig. 17.3 Pressure history at P1 for Case1

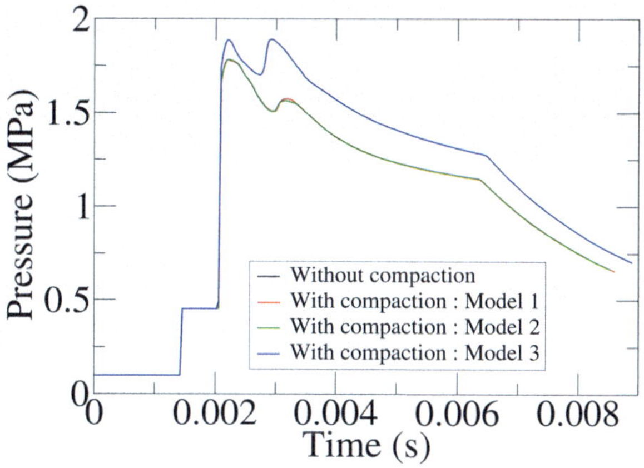

Fig. 17.4 Pressure history at P1 for Case3

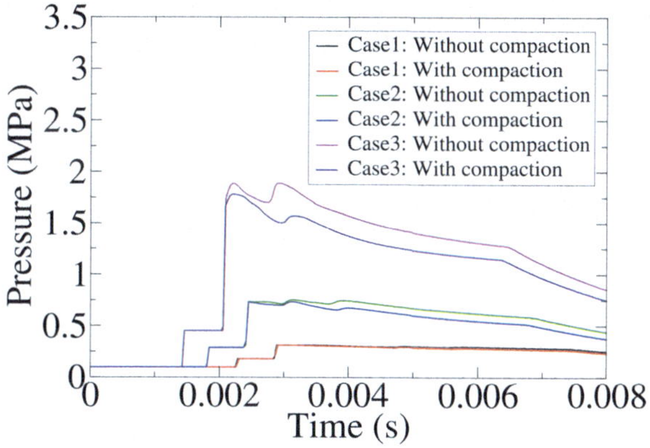

Fig. 17.5 Pressure history at P1

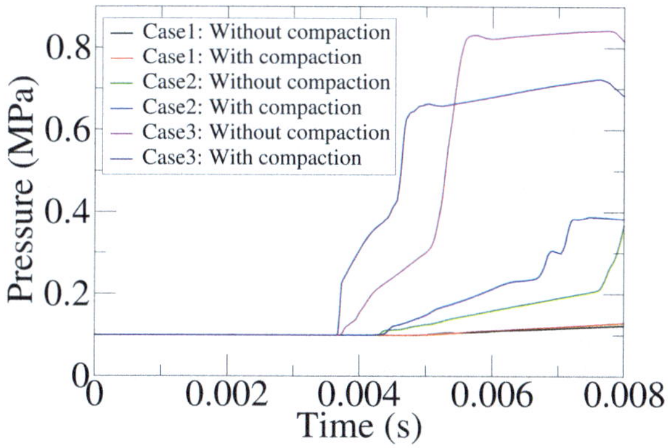

Fig. 17.6 Pressure history at P2

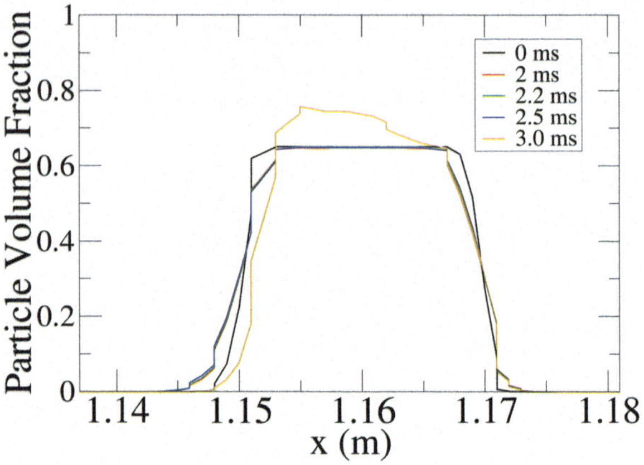

Fig. 17.7 Volume fraction for Case1 without compaction

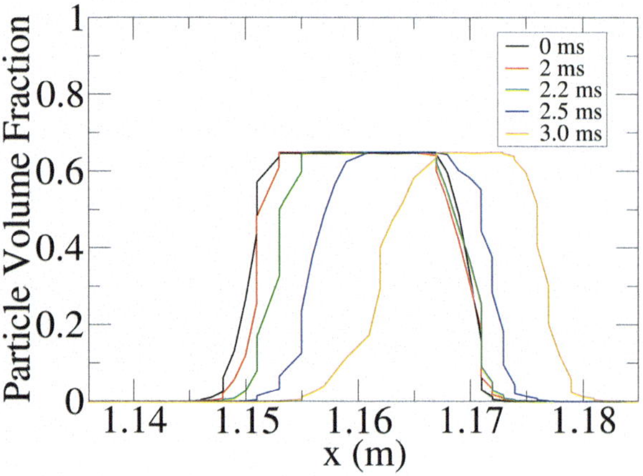

Fig. 17.8 Volume fraction for Case1 with compaction

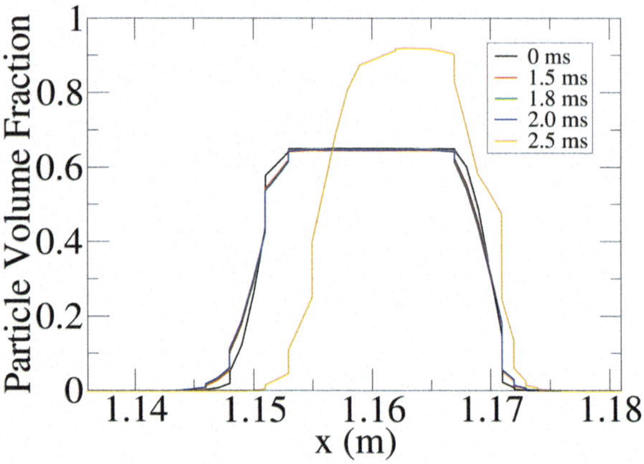

Fig. 17.9 Volume fraction for Case2 without compaction

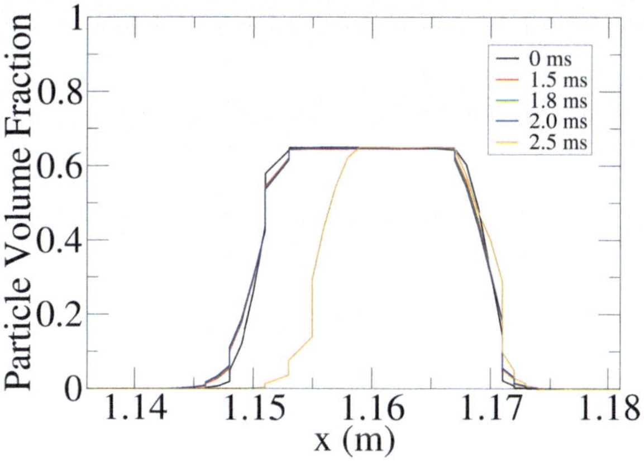

Fig. 17.10 Volume fraction for Case2 with compaction

17.6 Conclusion

A two-dimensional axisymmetric code was developed to study the interaction of shock on dense granular bed. The effect of compaction models was also carried out. The carry out the study, six equation model of the compressible multifluid formulation was used. The code was validated successfully. As shock transits the granular bed, the incident, reflected and transmitted shock was captured and was analyzed for various incident Mach numbers, particle diameter, particle volume fraction and compaction models.

References

1. X. Rogue, G. Rodriguez, J.F. Haas, R. Saurel, Experimental and numerical investigation of the shock-induced fluidization of a particles bed. Shock Waves **8**(1), 29–45 (1998)
2. Y. Ling, J.L. Wagner, S.J. Beresh, S.P. Kearney, S. Balachandar, Interaction of a planar shock wave with a dense particle curtain: modeling and experiments. Phys. Fluids **24**(11), 113301 (2012)
3. B.W. Skews, The reflected pressure field in the interaction of weak shock waves with a compressible foam. Shock Waves **1**(3), 205–211 (1991)
4. K.K. Kuo, J.H. Koo, T.R. Davis, G.R. Coates, Transient combustion in mobile gas-permeable propellants. Acta Astronaut. **3**(7–8), 573–591 (1976)
5. J.M. Powers, D.S. Stewart, H. Krier, Theory of two-phase detonation—part I: modeling. Combust. Flame **80**(3–4), 264–279 (1990)
6. K.A. Gonthier, J.M. Powers, A numerical investigation of transient detonation in granulated material. Shock Waves **6**(4), 183–195 (1996)
7. J.B. Bdzil, R. Menikoff, S.F. Son, A.K. Kapila, D.S. Stewart, Two-phase modeling of deflagration-to-detonation transition in granular materials: A critical examination of modeling issues. Phys. Fluids **11**(2), 378–402 (1999)
8. J.R. García-Cascales, J. Mulas-Pérez, H. Paillère, Extension of some numerical schemes to the analysis of gas and particle mixtures. Int. J. Numer. Meth. Fluids **56**(7), 845–875 (2008)

9. F. Zhen, L. Wang, Z. Wang, Numerical simulation of the deflagration to detonation transition in granular high-energy solid propellants. Cent. Eur. J. Energ. Mater. **16**(4) (2019)
10. R. Saurel, S. Le Martelot, R. Tosello, E. Lapébie, Symmetric model of compressible granular mixtures with permeable interfaces. Phys. Fluids **26**(12), 123304 (2014)
11. J.R. Gambino, A.K. Kapila, D.W. Schwendeman, Sensitivity of run-to-detonation distance in practical explosives. Combust. Theor. Model. **20**(6), 1088–1117 (2016)
12. T.P. McGrath, J.G. St Clair, S. Balachandar, A compressible two-phase model for dispersed particle flows with application from dense to dilute regimes. J. Appl. Phys. **119**(17), 174903 (2016)
13. P.T. Rao, *A Multi-Scale Approach for Modeling Shock Ignition and Burn of Granular HMX*, (Louisiana State University and Agricultural & Mechanical College, 2017)
14. C. López-Munoz, J.R. García-Cascales, F.J. Velasco, R.A. Otón-Martínez, An energetic model for detonation of granulated solid propellants. Energies **12**(23), 4459 (2019)
15. G.S. Yeom, K.S. Chang, A modified HLLC-type Riemann solver for the compressible six-equation two-fluid model. Comput. Fluids **10**(76), 86–104 (2013)
16. A.R. Elizabeth, T. Jayachandran, Numerical simulation of underwater propulsion using compressible multifluid formulation, in *Recent Asian Research on Thermal and Fluid Sciences* (Springer, Singapore, 2020), pp. 641–653
17. H. Paillere, C. Corre, J.G. Cascales, On the extension of the AUSM+ scheme to compressible two-fluid models. Comput. Fluids **32**(6), 891–916 (2003)
18. P. Batten, N. Clarke, C. Lambert, D.M. Causon, On the choice of wavespeeds for the HLLC Riemann solver. SIAM J. Sci. Comput. **18**(6), 1553–1570 (1997)

Chapter 18
Numerical Simulations of Shock/Boundary-Layer Interactions in an Axisymmetric Half-Isolator

Mohd Shahid Habib Khan, Jatinder Pal Singh Sandhu, and Santanu Ghosh

Abstract In this work Reynolds-averaged Navier-Stokes simulations of experiments performed on a cylindrical inlet by Funderburk and Narayanaswamy in AIAA J 57:1594–1607 (2019) are performed. Specifically, simulations of a Mach 2.5 flow with a fully turbulent flow model and a transition/turbulence model are computed and compared with the experimental data and numerical predictions from the aforementioned paper. It is observed that while the solution changes marginally with the use of a transition model, the results compare better with experimental data for the fully turbulent simulations, but using a shorter domain length.

18.1 Introduction

Shock-wave/Boundary Layer Interactions (SBLIs), as the name suggests occurs when a shock-wave interacts with the boundary layer developing on the flow surface. This phenomena can be observed in a wide range of flows such as high speed inlets, transonic/supersonic airfoils, nozzles etc. [1]. This shock-wave induces an adverse pressure gradient in the flow which leads to boundary layer thickening and if sufficiently strong can cause flow separation leading to pressure losses, increase in drag etc. [2–5].

Many studies have been performed to understand the physics of shock/boundary layer interactions and the issues caused by it. In most of these studies, the most commonly mentioned parameter for the shock induced separation is the length of the flow separation region, L_{sep}. This L_{sep} determines the severity of the shock-induced separation [2].

Most of the studies related shock-induced separation are performed for planar shock interactions with boundary layers developing on flat surfaces. Comparatively few studies have been done to study the effect of SBLIs on three-dimensional geometries [7–10]. There are applications such as engine inlets where SBLIs will occur on

M. S. H. Khan (✉) · J. P. S. Sandhu · S. Ghosh
Indian Institute of Technology, Madras, Chennai, India
e-mail: ae21s031@smail.iitm.ac.in

© The Author(s), under exclusive license to Springer Nature Singapore Pte Ltd. 2024
G. Rajesh et al. (eds.), *Advances in Shock Interactions*, Springer Proceedings in Physics 408, https://doi.org/10.1007/978-981-97-6099-2_18

negatively curved surfaces. Funderburk and Narayanswamy [6] performed one such study to analyze the effects of SBLIs on negative curvature geometries using an axisymmetric half-isolator (H-I) geometry. They used three different ramp angles (16°, 20° and 24°) representing different shock strengths and performed surface streakline visualisations and static pressure measurements. The results showed significantly increased values of L_{sep} and higher three-dimensionality for the H-I case compared to planar geometries. They also performed RANS computations on the 20° ramp angle case and their results showed qualitative agreement with the flow field in the experiments. In their computational study, the authors assumed the flow to be entirely turbulent, and performed their computations on a shortened domain compared to the experimental setup.

In this work, RANS computations have been performed on two different domains: domain A and domain B, matching the experimental and computational domains used in the work by Funderburk and Narayanswamy [6] respectively. While flow simulation on domain B is done with a fully turbulent flow assumption as done in [6], a transition model [13] is used for the simulation performed on domain A.

18.2 Methodology

All the numerical simulations in this work are performed for supersonic flow (Table 18.1) through a three-dimensional half-isolator (H-I) geometry [6] based on the the experimental setup (Fig. 18.1) of Funderburk and Narayanswamy [6] to study the effect of negative curvature on the mean structures of the shock/boundary-layer interactions (SBLI). The H-I geometry has a ramp that acts as a shock generator with a ramp angle $\alpha = 20°$ and $h/R = 0.246$ (h being height of the ramp and R being inner radius of the half-isolator). It is to be noted that the values of freestream velocity (u_∞) and Reynolds number (Re) calculated using the Mach number and free-stream pressure and temperature are a little different from the ones mentioned in the paper by Funderburk and Narayanaswamy [6] as listed in Table 18.1.

Table 18.1 Freestream parameters

Parameter	CFD (present)	Experiment [6]
M_∞	2.5	2.5
T_∞	140 K	140 K
p_∞	32.5 kPa	32.5 kPa
u_∞	593 m/s	583 m/s
Re	4.95×10^7 m^{-1}	5.3×10^7 m^{-1}

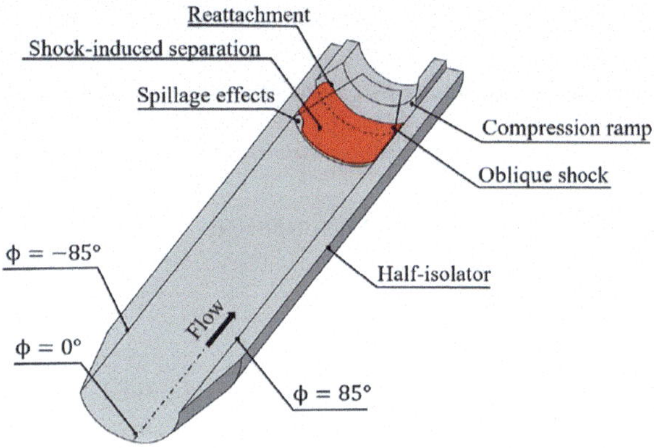

Fig. 18.1 Isometric view of H-I model [6]

18.2.1 Flow Solver

The numerical computations presented in this work are performed using an in-house finite-volume solver, FEST-3D [11]. It solves the three-dimensional Favre-averaged [16] Navier-Stokes equations on structured multi-block grids. In this work, a 3rd order accurate MUSCL [14] scheme was used for face reconstruction, AUSM [15] scheme was used for inviscid flux reconstruction while second-order central differences were used for viscous fluxes.

18.3 Computational Setup

Two domains having different streamwise extent were used in this study and are henceforward referred to as domain A and domain B. Domain A matches the streamwise extent of the experimental domain in [6] with an upstream extent of 380 mm from the leading edge of the ramp. The extents of this domain are as follows: $X = -380$ mm to $X = 50$ mm, $Y = 0$ mm to 95 mm and $\phi = 0°$ to $85°$ ($Z = 0$ mm to $Z = 50$ mm in Cartesian coordinate system). Here, $X = 0$ mm represents the leading edge (LE) of the compression ramp and $\phi = 0°$ represents the centerplane or the plane of symmetry. The grid of domain A has 2.12×10^7 cells. Flow simulation on domain A is performed using a one-equation transition model [13]. The motivation behind using a transition model is to better replicate the experimental study, wherein the boundary layer transitions from laminar to turbulent rather.

Domain B, on the other hand, matches with the streamwise extent of the computational domain used in [6], which has an upstream extent of 300 mm from the leading edge of the ramp. The extents of this domain is as follows: $X = -300$ mm to $X = 50$ mm, $Y = 0$ mm to 95 mm and $\phi = 0°$ to $85°$. The grid of domain B contains 0.853×10^7 cells. Here, X is along the streamwise direction, Y is along the vertical (wall normal) direction and ϕ is along the azimuth. A fully turbulent flow assumption is made for the simulation on this domain, as done in [6]; Menter's $k-\omega$ shear-stress transport model [12] is used for this case.

18.4 Results and Discussion

18.4.1 Inflow

The boundary layer (BL) data was obtained at the plane of symmetry ($\phi = 0°$) for all the domains and BL profiles are shown in Fig. 18.2. Measurement location is $X = -50$ mm for all the domains and a comparison has been made with the work by Funderburk and Narayaswamy [6]. The 98% BL thickness (δ) is 3 mm for Domain B and both the experimental data and RANS predictions of Funderburk and Narayanswamy [6] while for Domain A, it is around 3.75 mm. It can be observed that the BL profile predicted by the present computations on Domain B has a shape similar to that of the experimental profile [6], especially in the inner part of the boundary layer.

18.4.2 Flow Structure in Cross-Wise Planes

Normalised density contours are obtained using RANS simulations for three different locations, $X = 0$ mm (leading edge of the ramp) and $X = 30$ mm and $X = 35$ mm

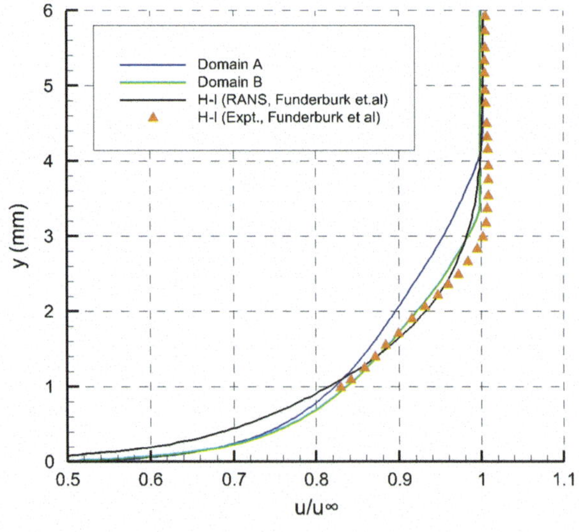

Fig. 18.2 Boundary layer profiles for the H-I models

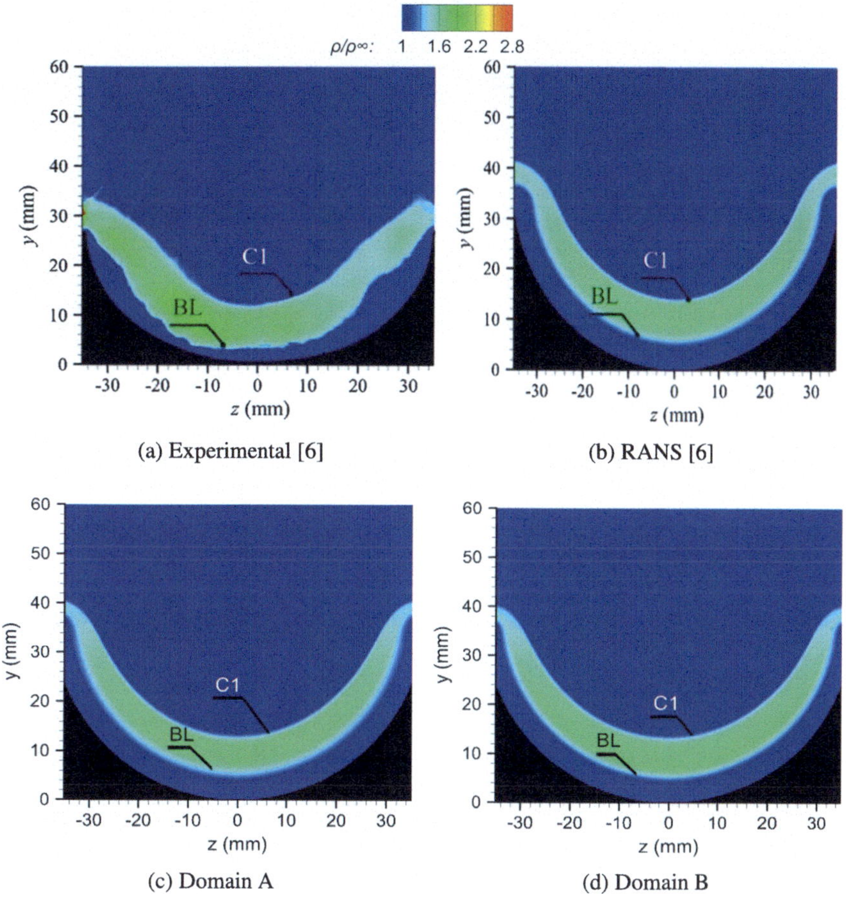

Fig. 18.3 Density contours on crosswise plane at X=0 mm (Leading edge of the ramp)

(30 mm and 35 mm downstream of the ramp leading edge respectively). These plots are compared with experimental and RANS contours presented in [6].

Figure 18.3 shows normalised density contours on a crosswise plane at $X = 0$ mm. A high density region can be observed bounded between an axisymmetric oblique shock (C1) and edge of separated boundary layer (BL). Also, the results obtained from the present computations look similar to the computations by Funderburk and Narayanaswamy [6], with the experimental data showing a thinner separated boundary layer.

Figure 18.4 shows density on a crosswise plane at X = 30 mm. Qualitatively, the RANS results from [6] and for Domain B (present computation) looks very similar to each other. However, in comparison, the experimental data shows a large core region and practically no boundary layer near the plane of symmetry, which is in stark contrast to the fully turbulent simulations and the simulation on domain *A* that

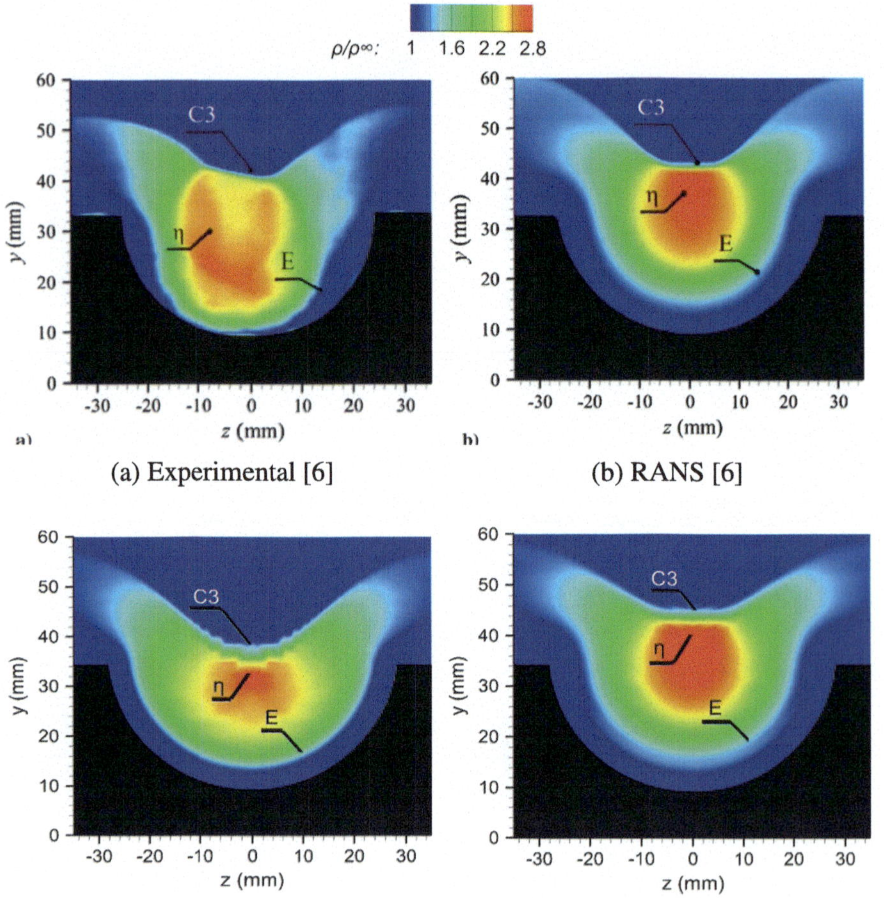

(a) Experimental [6]

(b) RANS [6]

(c) Domain A

(d) Domain B

Fig. 18.4 Density contours on crosswise plane at X = 35 mm (X = 30 mm for experimental case and Domain A)

considers flow transition. It is not apparent what is the cause of this difference, and has also not been discussed by the authors in the original paper[6]. However, it can be noted that simulation on Domain *A* (with transition model) produces the thinnest boundary layer at the symmetry plane among the computations.

18.4.3 Flow Structure in Streamwise Planes

Fig. 18.5 shows normalized pressure and streamwise velocity contours at centerplane ($\phi = 0°$). Figure 18.5a and b shows results from Funderburk and Narayanaswamy [6].

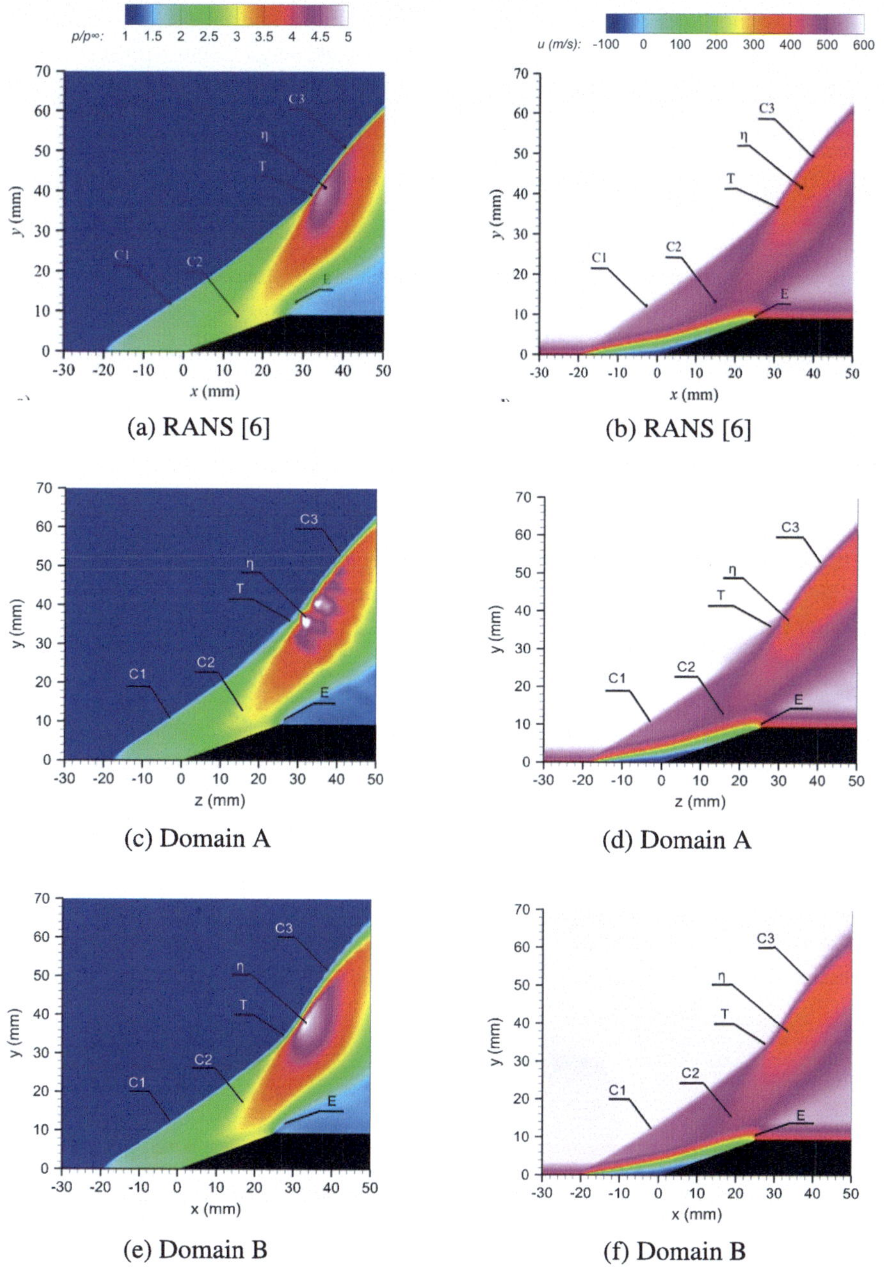

Fig. 18.5 Normalized static pressure contours (left) and streamwise velocity contours (right) at centerplane ($\phi = 0°$)

Figure 18.5c and d shows results for domain A while Fig. 18.5e and f shows results for domain B. The results look similar for all the cases although minor differences can be observed on closer examination. The maximum pressure ratio in the core region (η) is higher for both the present computations compared to RANS results from Funderburk and Narayanaswamy [6]. For the transition case (Domain A), the core region (η) and triple point of shock-shock interactions (T) is not clearly identifiable.

Streamwise velocity contours shows the flow separation region formed due to oblique-shock/boundary-layer interaction at the ramp. The incoming flow gets deflected by the separation bubble forming the separation shock (C1). A series of compression waves (C2) are formed due to area reduction in the flow. The resulting shock-shock interaction between C1 and C2 generates a triple point (T) and an outer shock (C3).

The length of the separation region normalised by the boundary layer thickness was given to be $L_{sep}/\delta \approx 11.7$ for the computations reported in [6]. In the present case, for domain B, separation starts at $X = -19$ mm and flow reattaches at $X = 16$ mm giving $L_{sep}/\delta \approx 11.7$, nearly equal to value reported in [6]. For domain A, flow separation starts at $X = -18$ mm resulting in $L_{sep}/\delta \approx 9$. It is to be noted that the BL thickness for domain A is higher compared to the other cases, which results in a smaller normalised separation length although the separation extents are similar to the other computations.

18.4.4 Effect of Transition Model

In order to check the effects of using a transition model on the development of a turbulent boundary layer, contours of intermittency have been plotted for the simulation on domain A. Figure 18.6 shows near wall intermittency (γ) contours for domain A near the inlet region. The value of intermittency is nearly one in the whole domain except very close to the wall, where it goes to zero. It can be observed that after some distance from the inlet, around X= -376 mm (or 376 mm upstream of the ramp

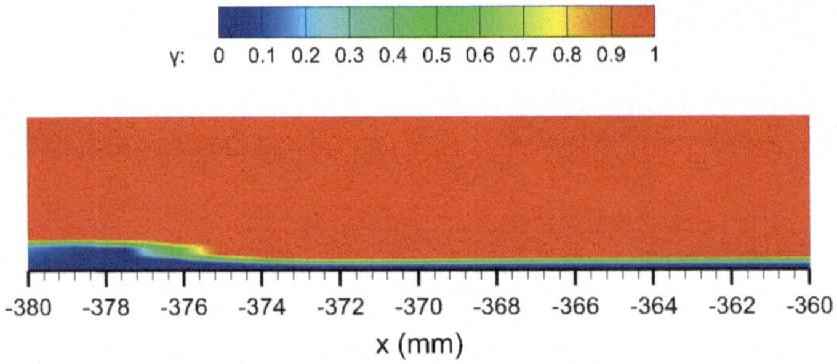

Fig. 18.6 Near wall intermittency (γ) contours at centerplane ($\phi = 0°$)

leading edge) in the streamwise direction, γ reaches a value of unity much closer to the wall, indicating that the boundary layer has transitioned into a turbulent one. It is apparent that the transition happens very early in the flow, which is suggestive of a bypass mode of transition. It is to be noted that free-stream turbulence intensity of 1% was used for this simulation.

18.5 Conclusions

RANS computations are performed on a negatively curved axisymmetric half-isolator geometry to study the shock/boundary layer interactions and the flow induced separation region formed at Mach no. of 2.5. Two different domain lengths based on the work by Funderburk and Narayanaswamy [6] were chosen and the obtained results are compared with the experiments from their work. Computations using a transition model for a full-length domain, and a fully turbulent flow assumption for a shorter domain were computed and contours of density, static pressure and streamwise velocity were obtained. It is observed that while the solution changes marginally with the use of a transition model, the results compare better with experimental data for the fully turbulent simulations, but using a shorter domain length.

Acknowledgements The authors would like to acknowledge Prof. Venkateswaran Narayanaswamy, North Carolina State University, for giving us permission to reuse the images from his publication.

References

1. N.T. Clemens, V. Narayanaswamy, Low-frequency unsteadiness of shock wave/turbulent boundary layer interactions. Ann. Rev. Fluid Mech. **46**, 469–492 (2014). https://doi.org/10.1146/annurev-fluid-010313-141346
2. H. Babinsky, J.K. Harvey (eds.), *Shock Wave-Boundary-Layer Interactions* (Cambridge University Press, 2011). https://doi.org/10.1017/CBO9780511842757. http://ebooks.cambridge.org/ref/id/CBO9780511842757
3. R.H.M. Giepman, F.F.J. Schrijer, B.W.V. Oudheusden, Flow control of an oblique shock wave reflection with micro-ramp vortex generators: effects of location and size, 066101(June 2014). https://doi.org/10.1063/1.4881941. http://dx.doi.org/10.1063/1.4881941
4. D. Dolling, Fluctuating loads in shock wave/turbulent boundary layer interaction: tutorial and update. (1993). https://doi.org/10.2514/6.1993-284
5. W. Huang, Z.B. Du, Yan, L., Xia, Z.X.: Supersonic mixing in air-breathing propulsion systems for hypersonic flights. Progress in Aerospace Sciences. https://doi.org/10.1016/j.paerosci.2019.05.005
6. L. Morgan, V. N. Funderburk, Investigation of negative surface curvature effects in axisymmetric shock/boundary-layer interaction, vol. 57, No. 4 (2019). https://doi.org/10.2514/1.J057275
7. M.L. Funderburk, V. Narayanaswamy, Experimental investigation of microramp control of an axisymmetric shock/boundary-layer interaction. AIAA J. **57**(8), 3379–3394 (2019). https://doi.org/10.2514/1.J057846

8. A. Zheltovodov, Shock wave/turbulent boundary layer interactions-fundamental studies and applications. in *Fluid Dynamics Conference, AIAA Paper 1996–1997* (1996). https://doi.org/10.2514/6.1996-1977
9. A. Zheltovodov, Some advances in research of shock wave turbulent boundary layer interactions, In 44th AIAA Aerospace Sciences Meeting and Exhibit, AIAA Paper 2006-0496, Jan. 2006. https://doi.org/10.2514/6.2006-496
10. F.J. Malo-Molina, D.V. Gaitonde, H.B. Ebrahimi, S.M. Ruffin, Three-dimensional analysis of a supersonic combustor coupled to innovative inward-turning inlets. AIAA J. **48**(3), 572–582 (2010). https://doi.org/10.2514/1.43646
11. J.P.S. Sandhu, A. Girdhar, R. Ramakrishnan, R. Teja, S. Ghosh, FEST-3D: finite-volume explicit STructured 3-Dimensional solver. J. Open Source Softw. **5**(46), 1555. https://doi.org/10.21105/joss.01555
12. F.R. Menter, M. Kuntz, R. Langtry, Ten years of industrial experience with the SST turbulence model. Turbulence, Heat and Mass Transfer 4, editor by K. Hanjalic, Y. Nagano, M. Tummers, Begell House, Inc., pp. 625–632 (2003)
13. F.R. Menter, P.E. Smirnov, L. Tao, R. Avancha, A one-equation local correlation-based transition model. Flow Trubulence Combust. https://doi.org/10.1007/s10494-015-9622-4
14. B. Van Leer, Towards the Ultimate Conservative Difference Scheme. V. A Second-order Sequel to Godunov's Method. J. Comput. Phys. **32**(1), 101–136 (1979). https://doi.org/10.1016/0021-9991(77)90095-X
15. M.-S. Liou, C. Steffen, A new flux splitting scheme. J. Comput. Phys. **107**, 23–39 (1993)
16. D. Wilcox, Effects of compressibility, in Turbulence Modeling for CFD, 3rd Edn, Vol. 1, DCW Industries (2006) pp. 239–297

Chapter 19
Effect of 'Ramped Vane' Vortex Generators on a Flare-Induced Shock—Boundary Layer Interaction

T. Nilavarasan, G. N. Joshi, A. Misra, C. Manisankar, and S. B. Verma

Abstract This paper describes the modifications induced by Ramped Vanes (RVs) on an axisymmetric compression corner-induced shock—boundary layer interaction flowfield. Experimental and numerical investigations were performed at Mach 2 over a cone-cylinder-flare model, which had flow turning angle of 24° at the cylinder-flare juncture. The RVs were located 50 mm upstream of the corner and had a height (h) equivalent to 50% of the local boundary layer thickness ($h = 0.5\delta_{RV} = 2.1$ mm). The counter-rotating vortices emanating from these RVs formed alternate upwash and downwash bands in the upstream boundary layer, which led to premature and delayed onset of flow separation respectively. Additionally, interaction of these vortices with the recirculation region, caused formed tornado-like vortices and brought the reattachment line very close to the compression corner. Along the downwash region, the separation shock was also considerably pushed downstream, which increased its oscillating frequency range, while roughly maintaining the same Strouhal number range.

19.1 Introduction

Shock—Boundary Layer Interactions (SBLIs) are quite common during the atmospheric flight phase of a launch vehicle. These interactions are known to trigger an aero-elastic phenomenon called "buffeting" [1], which might lead to bending/shell-breathing vibrations of the vehicle structure. These are extremely hazardous at transonic/low-supersonic Mach numbers and in high dynamic pressure conditions, as the amplitude of the oscillating aerodynamic loads could be fairly larger under those situations [2]. Furthermore, the frequency range of these shock oscillations often overlap with the natural frequency range of the skin panels [3], enhancing the

T. Nilavarasan (✉) · G. N. Joshi · A. Misra
Defence Institute of Advanced Technology, Pune, Maharashtra 411025, India
e-mail: nlvrsnt@gmail.com

C. Manisankar · S. B. Verma
National Aerospace Laboratories, Bengaluru, Karnataka 560017, India

© The Author(s), under exclusive license to Springer Nature Singapore Pte Ltd. 2024
G. Rajesh et al. (eds.), *Advances in Shock Interactions*, Springer Proceedings in Physics 408, https://doi.org/10.1007/978-981-97-6099-2_19

chance of resonance. Hence, it is highly imperative to develop workable flow control methods that can manipulate these interactions and ensure safe operation of these vehicles.

One of the prominent reasons for SBLIs to occur on launch vehicles, is the presence of inter-stage flares. At supersonic speeds, these sections act as axisymmetric compression corners and turn the incoming flow away from the vehicle centerline through a conical shock wave. If this shock imposes a sufficiently strong adverse pressure gradient, a separation region is formed around the compression corner. Many studies [4–8] have been conducted in the past to understand flare-induced SBLIs. While the interactions are qualitatively similar to those induced by ramps [9], the three-dimensional relieving effect [10] makes the axisymmetric interactions relatively weaker and smaller for the same inflow condition and flow deflection angle [11]. Past research on compression ramps have shown that the separation and reattachment shocks undergo low-frequency oscillations, due to the periodic 'growth and collapse' of the separation bubble [12]. A study by Verma and Koppenwallner [13] on a flare-induced interaction, also confirmed these observations. Furthermore, Erengil and Dolling's [14] experiment with a swept compression corner indicated that the separation length and the shock frequency could be inversely related. Hence, there is a growing interest in developing flow control methods that will attenuate the shock-induced flow separation and thereby push the shock oscillations to higher frequencies that are less detrimental to the vehicle.

Many flow control techniques have been proposed [15], among which the installation of mechanical vortex generators [16] showed great promise, as they do not require additional power to operate and are relatively simpler to design. These are either wedge/ramp shaped or vane-shaped protuberances that are installed ahead of the interaction region, to produce streamwise vortices that can entrain high-momentum fluid into the near-wall region and thereby enable the boundary layer to push through the adverse pressure gradient more effectively. Other novel shapes have also been developed and compared [17–19] among which, there has been considerable interest on Ramped Vanes (RVs) [20–22], as they perfectly combine the sturdiness of a ramp/wedge and the separation control performance of the vanes.

In the present study, a circumferential array of RVs were employed ahead of an axisymmetric compression corner in supersonic flow, to assess their influence on the shock—boundary layer interaction flowfield. Numerical simulations were performed to bring out the topological changes in the separation region. Also, real-time wall pressure measurements were obtained from wind tunnel tests, through discretely located Kulite transducers, to analyze the modifications induced in the unsteady shock motions.

19.2 Wind Tunnel Testing

19.2.1 Model and Inflow Details

Experiments were carried out in the 0.3 m Trisonic wind tunnel, located at CSIR-NAL, Bengaluru. The test model had a cone-cylinder-flare configuration (Fig. 19.1), with a flow turning angle of 24° at the cylinder-flare juncture. It was mounted along the tunnel centerline, with zero angle of attack and zero side-slip angle (Fig. 19.2). The forebody length (L) ahead of the compression corner was 380 mm, which along with the cone's semi-angle corresponds to the maximum boundary layer thickness achievable at the corner, while averting the interference of the apex shock reflection from the tunnel walls. Simulations conducted in the absence of the flare (cone-cylinder configuration) estimated this boundary layer thickness (δ) to be 5 mm. Details on the model's construction have been presented in Ref. [22].

A 40 mm long and 5 mm deep groove was made at 50 mm upstream of the corner, to install the interchangeable RV inserts on their exterior surfaces. The trailing edges of the RVs aligned perfectly with those of the inserts and hence the overall distance

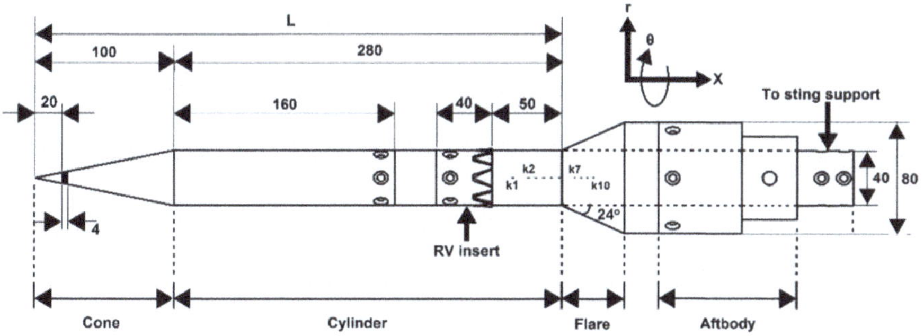

Fig. 19.1 Schematic diagram of the wind tunnel model

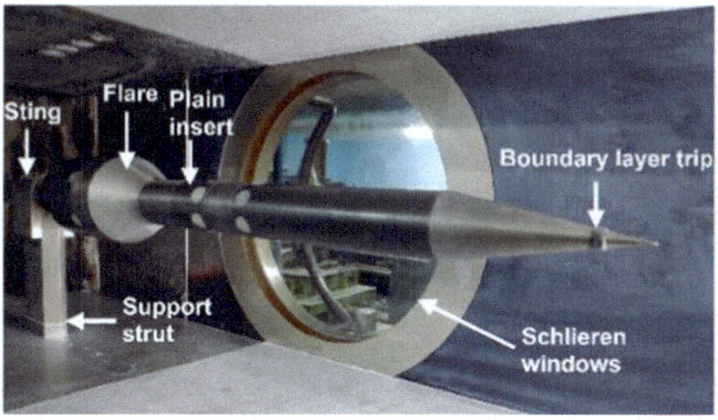

Fig. 19.2 Photograph of the model mounted inside the wind tunnel

between these devices and the corner was 50 mm (10δ). This location was chosen because, an earlier study on axisymmetric compression corners [23] had shown 10δ to be the optimal streamwise distance between the vortex generators and the compression corner, for achieving maximum separation control. The local boundary layer thickness (δ_{RV}) at this location was found to 4.2 mm. Plain inserts that had no vortex generators on their exterior surface was used to study the uncontrolled interaction (Fig. 19.2).

At 50 mm upstream of the corner, a 40 mm long and 5 mm deep groove was cut to enable installation of interchangeable inserts that contained the RVs on their exterior surfaces. The trailing edges of the RVs aligned perfectly with those of the inserts and hence the overall distance between these devices and the corner was 50 mm (10δ). This location was chosen because, an earlier study on axisymmetric compression corners [23] had shown 10δ to be the optimal streamwise distance between the vortex generators and the compression corner. The local boundary layer thickness (δ_{RV}) at this location was found to 4.2 mm. Plain inserts that had no vortex generators on their exterior surface was used to study the uncontrolled interaction (Fig. 19.2).

The geometry of RVs was adopted from the work of Lee et al. [17] and displayed in Fig. 19.3a, where it can be seen that all of its dimensions are based on the device height (h). Based on Anderson et al. [24], it was decided to have an inter-device spacing of 7.5 h between the devices (Fig. 19.3b) and hence the circumferential arrangement will be as per the following simple arithmetic expression:

$$\pi * d = n * (7.5h) \tag{19.1}$$

Here, the diameter (d) of the cylinder is fixed as 40 mm (Fig. 19.2) and ‘n’ is the total number of devices in the array, which can only be an integer. In the present study, it was decided to have a total of 8 devices in the array, which corresponds to a device height (h) of 2.1 mm ($h = 0.5\delta_{RV}$). The other dimensions and the inter-device spacing were computed as per the details provided in Fig. 19.3.

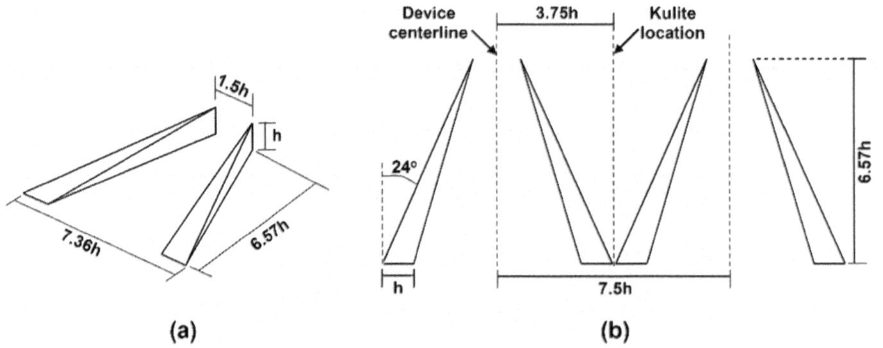

Fig. 19.3 Ramped Vanes (RVs). **a** Isometric view of a single device; **b** Planform view showing the location of the Kulites

19.2.2 Pressure Measurements and Digital Data Acquisition

A Druck 4010 series transducer monitored the total pressure (P_O) at the settling chamber. It had a range of 1379 kPa (200 psia) and a full scale accuracy of $\pm$ 0.1%. The freestream pressure (P_{Inf}) at the test section was recorded using an ESP scanner from Pressure Systems Inc, which had a range and full-scale accuracy of 207 kPa (30 psia) and 0.04% respectively. After conditioning, the analog signal was digitized using a 16-bit A/D converter (NI6036). For this experiment, 500 samples were recorded at a rate of 500 Hz, which amounts to an overall acquisition time of 1 s. The total and freestream pressure (P_O and P_{Inf}) had an uncertainty of $\pm$ 1.4 kPa and $\pm$ 0.7 kPa respectively.

The instantaneous pressure measurements were obtained using 10 Kulite transducers (Model No: XCQ-093). These transducers were 9.5 mm long and had a diameter of 2.4 mm. They had a range of 172 kPa (25 psia) and a sensitivity of 0.4–0.6 mV/kPa. The protective M-screen that was provided over the sensing element limited the frequency response of these transducers to 50 kHz. They were arranged in a straight line (labelled k1 to k10 in Fig. 19.4a) along the gap between two adjacent devices, because the computations showed maximum delay in flow separation along this region. The transducers were placed inside recess cavities of 9 mm length and 2.5 mm diameter, to further protect them from any contamination in the incoming flow. The sensing elements are connected to the exterior surface through a small orifice, which was 1 mm long and had a diameter of 0.5 mm. This cavity had a resonant frequency of about 55.43 kHz. This is not problematic, as the acquisition was carried out at 50 kHz and a low pass filter of 20 kHz was applied before processing. The time delay due to this arrangement was less than 1 μs [25]. The streamwise position of all 10 Kulites have been graphically represented in Fig. 19.4b.

A NI4495 DC series data acquisition card with a 24-bit resolution was used to record the pressure signals. The Kulites were excited using a 9 V DC supply and their output signals were amplified by a gain of 10. A total of 200 records were acquired by each transducer consecutively, with 4096 data points per record. Thus 819,200 data points were recorded by every transducer, amounting to an acquisition time of

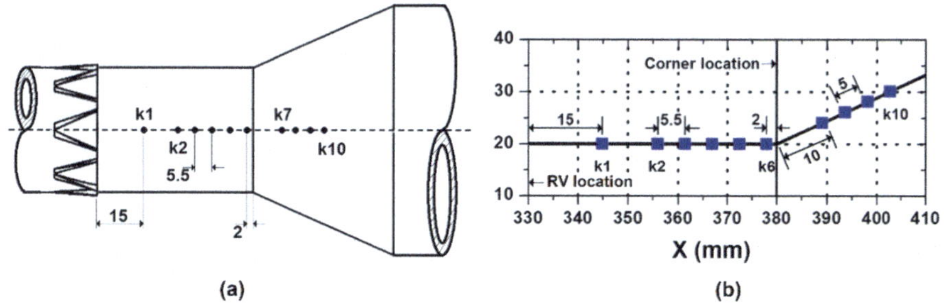

Fig. 19.4 **a** Top view of the wind tunnel model near the region of interest; **b** Graph showing the streamwise position of the Kulites

about 16 s. The spectral analysis of these signals were performed through Welch's method [26], with a frequency resolution of 12.2 Hz.

Before they were installed into the cavities, the Kulites were subjected to a seven point calibration procedure, using a Druck calibrator (Model No: DPI-610). The calibration was performed statically, as variations between the static and dynamic calibrations are very low [27]. A single point check calibration was done every day until the experiments were completed to rule out any deviations in the calibration constants. The uncertainty in the calibrations was found to be within $\pm$ 1% of the full-scale value. Also, the repeatability of the peak standard deviation was within $\pm$ 0.03 kPa.

19.2.3 Schlieren Imaging

The flowfield was visualized using a conventional Z-type Schlieren arrangement. The circular windows (diameter = 0.28 m) on either side of the test section provided optical access (seen in Fig. 19.2). An IL-106G LED illuminator was used as a light source, which emitted non-coherent green light (wavelength = 523 nm to 535 nm) in pulse mode. Two spherical mirrors with a diameter of 0.3 m and a focal length of 3.0 m, were arranged 6.0 m (twice the focal length) apart from each other on either side of the test section, to collimate and refocus the light rays. A horizontal knife-edge was employed at the focal point, as it clearly distinguished the upstream boundary layer. Fifty frames were recorded using a Motion Pro X3 camera at 60 fps (Total acquisition time = 0.83 s), with an exposure time of 125 μs and an image resolution of 3.2 pixels/mm.

19.3 Computational Approach

19.3.1 Solver Setup

The simulations were performed in STARCCM+. A steady, density-based Algebraic Multi-Grid (AMG) solver was used, in a coupled-implicit approach. The Advection Upwind Splitting Method (AUSM) [28] was used to evaluate the convection and diffusion fluxes, whose discretization was second order accurate in space. The Sutherland law was employed to determine the dynamic viscosity of air, which was assumed to be an ideal gas. The Spalart–Allmaras model [29] was used for turbulence closure.

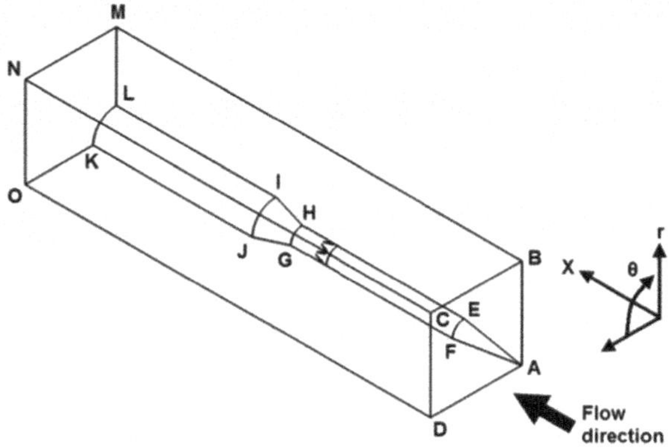

Fig. 19.5 Wireframe view of the computational domain

19.3.2 *Flow Domain*

The same model geometry and flow conditions from the wind tunnel tests were used for the numerical simulations, so that the exact flowfield can be reproduced. But due to the axisymmetric nature of the flowfield, only one quarter of it was simulated, as shown in Fig. 19.5. The dimensions of the cuboidal domain was 150 mm × 150 mm × 700 mm. The model axis was superimposed over one of the longitudinal axes of the domain, making the surfaces ABMLIHE and ADOKJGF as symmetry planes. The freestream flow conditions were provided at the boundaries ABCD, BCNM and CDON, while the surface KLMNO was designated as a pressure outlet. The model's solid boundaries were all assumed to be adiabatic walls with a no-slip condition.

19.3.3 *Meshing*

The flow domain was discretized into a hybrid mesh that mostly contained hexahedral cells. As seen in Fig. 19.6a, a dense surface mesh (maximum cell size = 0.75 mm) was provided near at the region of interest. When the RVs were present, an even denser annular volumetric mesh was generated around them (maximum cell size = 0.3 mm), to predict the intricate flow features in their vicinity (Fig. 19.6b). Cross-sectional views of the meshes (from XX' in Fig. 19.6) have been presented in Fig. 19.7. Fifty prismatic layers were provided in the near-wall region (within 5 mm from the surface) at a growth rate of 1.2 and a minimum cell wall distance of $r^+ = 0.01$, to estimate the boundary layer flow (Fig. 19.7a). Given that the geometrical features of the RVs were less than 5 mm, the prism layers over them were retracted, with a minimum overall thickness of 0.05 mm (1% of the regular thickness). Nevertheless, the dense annular mesh provided adequate mesh resolution in the regions immediately above the RVs (Fig. 19.7b). The cells in the interior parts of the domain had a uniform

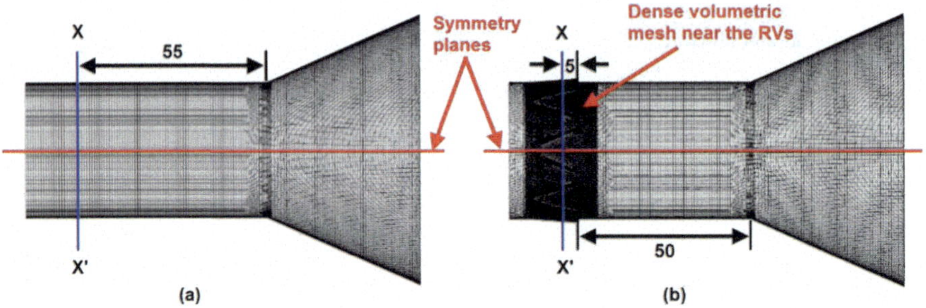

Fig. 19.6 Surface mesh near the compression corner. **a** Uncontrolled; **b** with Ramped Vanes (RVs)

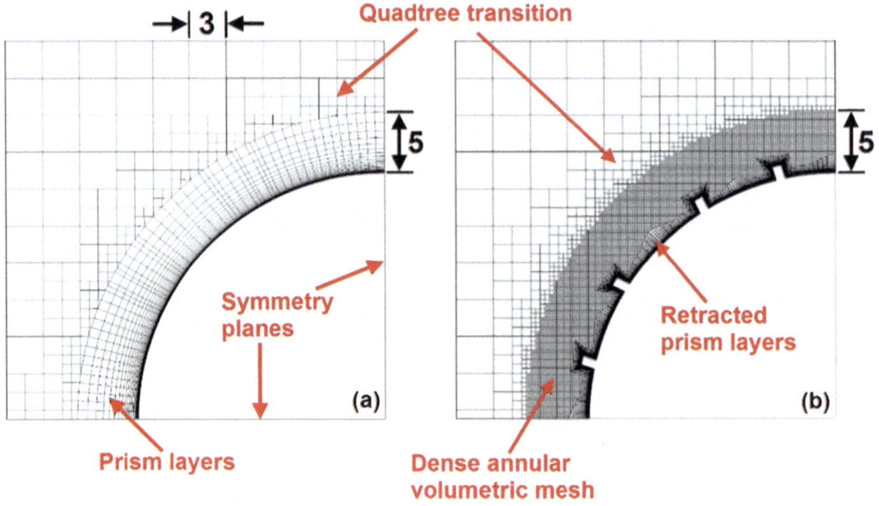

Fig. 19.7 Cross-sections of the meshes at XX'. **a** Uncontrolled; **b** Ramped Vanes

size of 3 mm and a quad-tree transition was provided between them and the densely packed cells in the near-wall region.

19.3.4 Code Validation

It has been previously established that this computational methodology can accurately predict axisymmetric [23] compression corner flows. Also, it has been demonstrated that the setup is capable of accurately determining the flow behavior in the regions close to the vortex generator devices [23]. Nevertheless, the Kulite data was time-averaged and compared with the computational pressure distributions, to

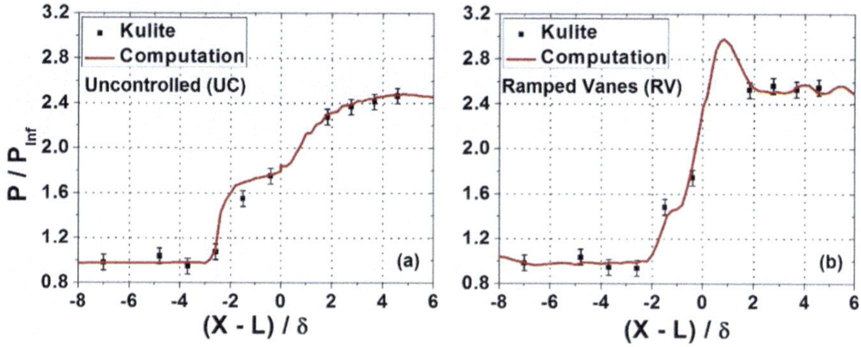

Fig. 19.8 Comparison of experimental and computational mean pressure distributions. **a** Uncontrolled; **b** Ramped Vanes

ensure reliability. As such, the plots presented in Figs. 19.8a, b show that the numerical results closely align with the experimental data. It is also important to note that the computational pressure distribution in Fig. 19.8b was obtained from the region along the gap between two adjacent RV devices, matching the location of the Kulite transducers.

19.4 Results

19.4.1 Schlieren Visualization

The time-averaged Schlieren images are presented in Fig. 19.9. The Mach waves originating from the joints in the model do not cause any perturbations in the interaction flowfield. Similarly, the reflection of the apex shock also impinges just beyond the flare's shoulder and hence does not affect the flow in the region of interest.

The distance between the projected impingement location of the separation shock's foot and the compression corner is termed as the interaction length (X_{IL}). For the uncontrolled interaction (Fig. 19.9a), this distance was found to be 12.5 mm (2.5δ). As the RVs were brought in (Fig. 19.9b), the interaction length (X_{IL}) was reduced by about 39.2%. Besides, the RVs also induced the formation of compression (C) and recompression (R) shocks, with the angle between them being 6°. Moreover, the reattaching boundary layer appeared to be considerably thicker with the RVs, which could possibly be due to the counter-rotating vortices that emerge from them. The main flare shock had an inclination of about 41°, which perfectly matched with the theoretical estimate of 40.86° from the conical shock relations [10].

The other important features in Fig. 19.1 such as the triple point and the reattachment shock are not clearly distinguishable in Fig. 19.9, partly because of the smaller interaction scale and partly because of the three-dimensional nature of the flowfield.

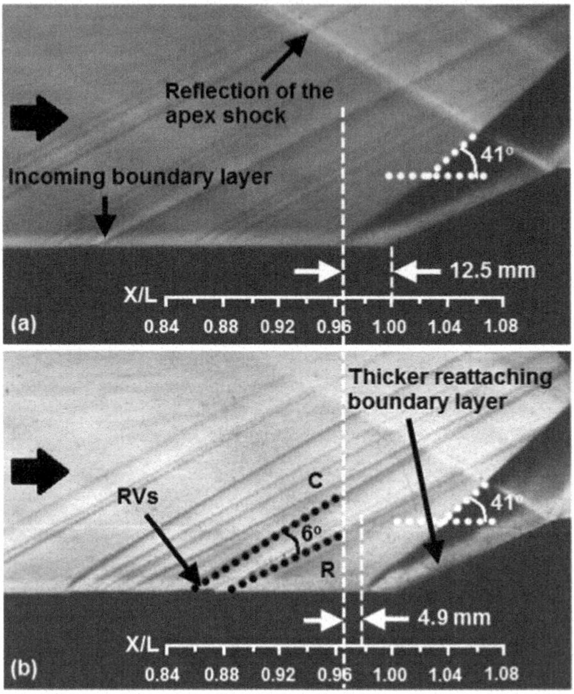

Fig. 19.9 Time-averaged Schlieren photographs. **a** Uncontrolled; **b** Ramped Vanes

Given that the Kulites were facing the tunnel ceiling, it is plausible that the interaction length (X_{IL}) measured from Fig. 19.9b largely corresponds the device gap region. Also the snapshots were recorded at a low sampling rate of 60 fps. Hence, these observations are merely anecdotal and more detailed three-dimensional characterizations of the flowfield are presented through the numerical results in the following section.

19.4.2 Separated Flow Topology

Figure 19.10 presents the surface pressure distribution and streamline patterns. In the absence of RVs, the pressure rise due to the interaction began well ahead of the compression corner and was roughly uniform across the entire circumference of the model (Fig. 19.10a). The average distance between the corner and the location of initial pressure rise (upstream influence) and was found to be 14 mm (2.8δ) for the uncontrolled case. After the initial jump across the upstream influence line (marked as a dark line in Fig. 19.10), there was very little change in pressure until the compression corner. This 'pressure plateau' region denotes the presence of flow separation, which could be very clearly visualized in Fig. 19.10b. The separation line on the cylinder surface had minor undulations, while the reattachment line on the flare surface was highly corrugated. Also, striation patterns could be observed beyond the reattachment line. These secondary three-dimensional effects are likely due to the presence of Goertler vortices, which has been discussed in detail in Ref. [22]. More importantly, the average distance between the separation line and the corner was 12 mm (2.4δ), which matched with the surface oil flow visualizations (also presented in Ref. [22]).

As the RVs were introduced (Figs. 19.10c, d), the pressure distribution at the interaction region displayed a sinusoidal pattern (Fig. 19.10c), with crests along the device centerline and troughs along the device gap regions. Additionally, the contour also indicated that the increase in pressure along the device centerline (within the crest) happened more gradually, while the same along the device gap region (downstream of the trough) appeared to be much more rapid. Besides, high pressure spots could be noticed along the device gap region, immediately downstream of the compression corner. Thus, it is inferred that in addition to modifying the interaction scale, the RVs had considerably altered the smearing of both the separation and reattachment shocks. In Fig. 19.10d, the streamlines downstream of the RVs could be seen converging along the device centerlines, indicating the formation of counter-rotating vortices. The separation region had broken up into large spade shaped regions, which had their apexes aligned along the device centerline. The reversed flow regions appeared to

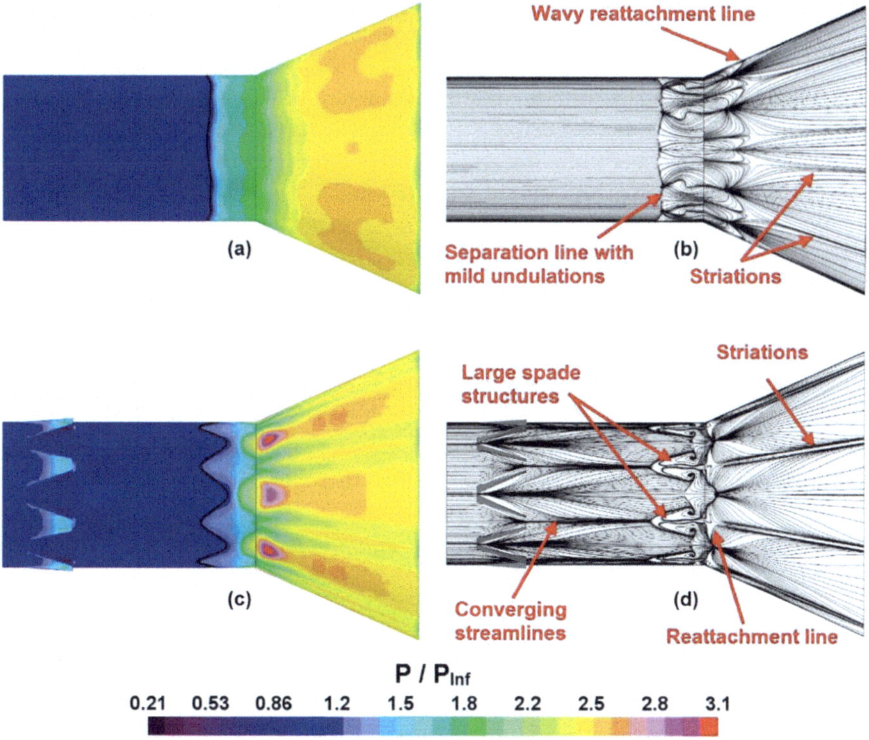

Fig. 19.10 Surface pressure distributions and streamlines. **a, b** Uncontrolled; **c, d** Ramped Vanes

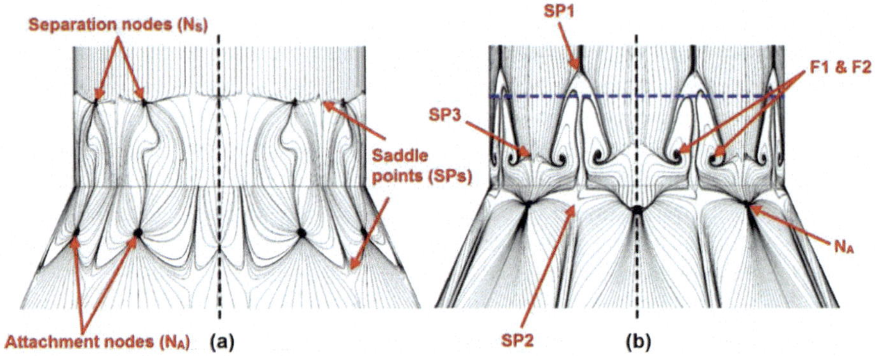

Fig. 19.11 Critical points in the separated flow region **a** Uncontrolled; **b** Ramped Vanes

have greatly reduced between these spade patterns (along the device gaps). On the flare surface, the reattachment line had moved very close to the compression corner. Both the corrugations on the reattachment line and the striations on the flare were well-organized and could be easily correlated with the RV arrangement.

Figure 19.11 displays a closer view of the separated flow regions. In the uncontrolled case (Fig. 19.11a), both the separation and reattachment lines were made up of a series of node-saddle point arrangements, similar to the model presented by Lighthill [30]. The nodes were situated closer to the compression corner than the saddle points, giving rise to the corrugated nature of these lines. As mentioned earlier, the undulations were considerably stronger in the reattachment line.

The modified separated flow region is shown in Fig. 19.11b. The spade patterns appeared to house two foci type singularities (F1 and F2) in their lobes, denoting that the interaction between the counter-rotating vortices and the reversed flow region had induced the formation of tornado-like vortices. The apex of the spade was a saddle point (SP1). The reattachment line still retained an attachment node-saddle point arrangement, but as discussed above, their organization was closely related to the RV array (SP2 along the device centerline and N_A along the device gap region). Just upstream of N_A and on the cylinder surface, a third saddle point (SP3) could also be noticed. When the attachment and separation node formed at the foremost and rearmost ends of the model are taken into account (not visualized here), the topology agrees with the summation rule provided by the critical point theory [31].

$$\sum(Nodes + Foci) - \sum(Saddlepoints) = 2 \tag{19.2}$$

19.4.3 Separation Shock's Unsteadiness

The Kulites (k1–k10) obtained real-time wall pressure measurements along the device gap region, where the separation shock appears to have been pushed downstream towards the corner. The standard deviation (σ) of the acquired signals have been presented in Fig. 19.12. During the uncontrolled interaction, a sharp rise in 'σ' was observed at k4, denoting that the separation shock might have been oscillating across this transducer. As the RVs were installed, this sharp peak had disappeared, indicating that the shock-intermittent region had moved away from k4. Based on the visualizations presented in Figs. 19.9, 19.10 and 19.11, it can be deciphered that the separation shock must have moved downstream towards the corner.

The signals from k4 were evaluated statistically, by constructing their Probability Density Functions (PDFs). Figure 19.13 compares these PDFs, alongside the Gaussian distribution obtained from k1, which is well upstream of the interaction. Both PDFs from k4 were positively skewed. However, the PDF from the uncontrolled interaction had a significantly thicker tail, denoting a higher probability of finding data points that carry the pressure downstream of the separation shock. Contrarily, the signal acquired in the presence of RVs, showed a very high modal peak just below the mean value, signifying that most of its data points carried the pressure beneath the undisturbed boundary layer and that the shock very rarely moved ahead of the transducer. This clearly establishes that the RVs pushed the shock-intermittent region downstream.

The power spectra of the signals from k4 are plotted in Fig. 19.14. The frequency content in the signal is preserved throughout the shock intermittent region [32], but the area beneath each spectrum is directly proportional to the variance (σ^2) of the corresponding signal. Hence, each spectra (G(f).f) was divided by σ^2, to facilitate comparison. For the uncontrolled interaction, the dominant frequency range was found to be 0.55 kHz to 0.9 kHz. In the presence of RVs, the same had shifted to 1.1 kHz to 1.7 kHz (Fig. 19.14a). The Strouhal number (Str) was computed based on the

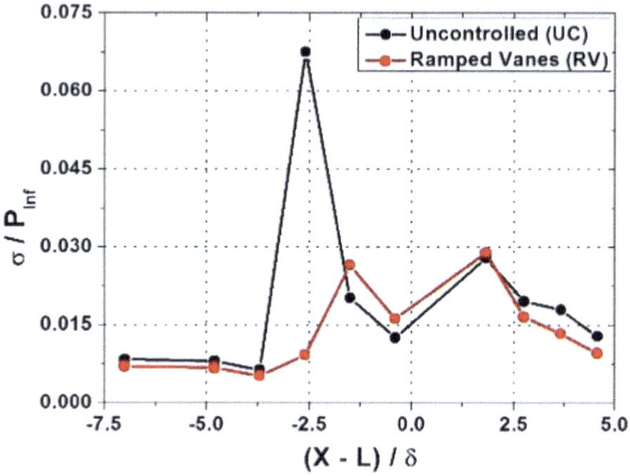

Fig. 19.12 Standard deviation distributions

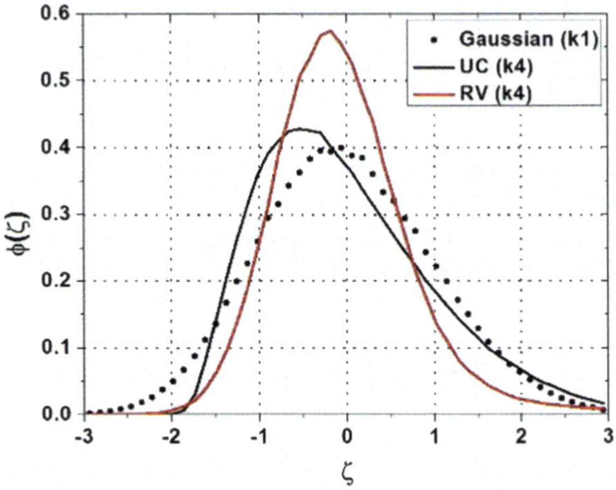

Fig. 19.13 Probability density functions

interaction length (X_{IL} from Fig. 19.9), to express the frequency non-dimensionally.

$$Str = \frac{f.X_{IL}}{V_{Inf}} \tag{19.3}$$

When the same spectra was plotted based on the Strouhal number Fig. 19.14b, the shock oscillations in the uncontrolled interaction was in the range of 0.013–0.022. Similarly, the range was 0.015–0.025 when the RVs were present, denoting no major change in the Strouhal number.

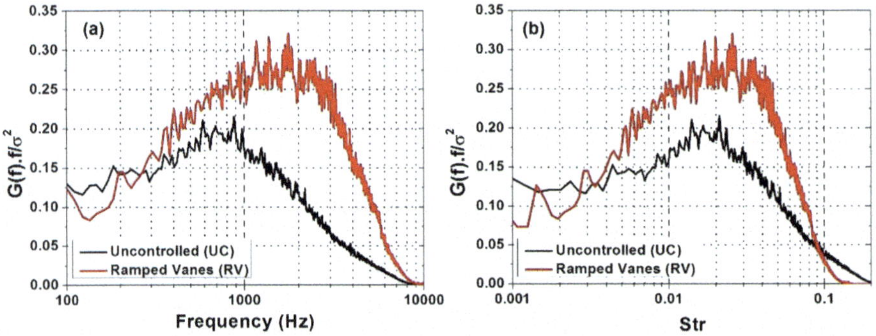

Fig. 19.14 Normalized power spectral densities from k4 in terms of **a** Frequency (f) and **b** Strouhal number (Str)

19.4.4 Vortex Development and Boundary Layer Entrainment

The changes observed in the interaction region are caused by counter-rotating vortices that emerge due to pressure difference induced across the span of each RV device. The converging passage between the vane pair caused a significant increase in static pressure (Fig. 19.15a). Furthermore, the flow turning induced by the ramped leading-edges also caused a minor increase in pressure. Concurrently, flow expanded through the diverging passage along the device gap region. Thus, the high pressure fluid from around the device centerline, was forced to roll across the vanes to reach the device gap region. In this process, low-pressure vortex regions were formed, right next to the leeward side of the vanes. The vortical flow regions can be recognized in Fig. 19.15b, as the locations where the streamlines having a strong azimuthal component. The expansion along the device gap region was terminated by glancing shocks that originated from the RV trailing edges on either side. These shocks also turned the vortical flows and realigned them along the streamwise direction. The central flow emerging through the trailing edge gap, kept the vortices apart and reduced interaction between them. This is evident from the streamlines converging into two distinct lines (one on either side of the device centerline) downstream of the trailing edges. Past studies [19, 20] that compared different vortex generators have attributed the superior separation control performance of RVs to this separation of vortices along the device centerline.

The cross-sectional contour in Fig. 19.16a shows the normalized vorticity distribution at 20 mm upstream of the corner (just ahead of the interaction). The plot confirms the existence of counter-rotating vortical structures that are responsible for the alterations observed in the interaction region. Likewise, Fig. 19.16b depicts a parameter called 'relative velocity (V_R), which is calculated by subtracting the streamwise velocity field from the uncontrolled interaction ($V_{X(UC)}$) from that of the

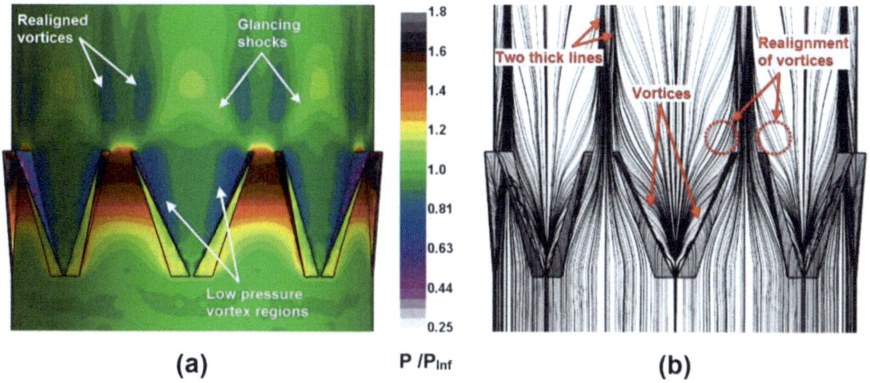

Fig. 19.15 Flowfield in the vicinity of the RVs. **a** Surface pressure distributions; **b** Surface streamline patterns

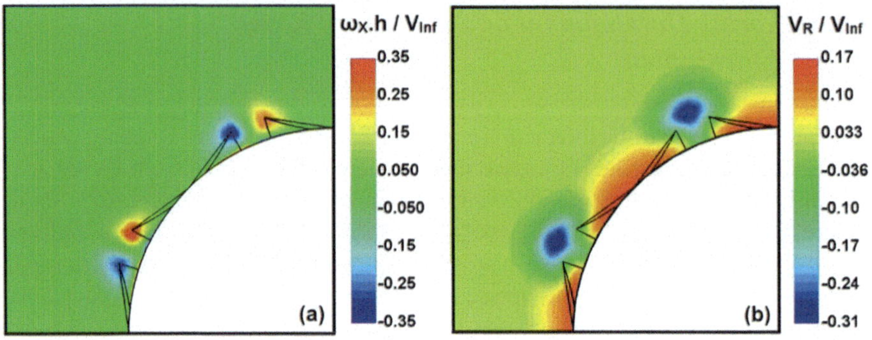

Fig. 19.16 Cross-sectional visualizations from 20 mm (4δ) upstream of the corner. **a** Vorticity; **b** Relative velocity

controlled interaction (V_X).

$$V_R = V_X - V_{X(UC)} \tag{19.4}$$

Because of the vortical action, low momentum fluid from the near-wall region was huddled close to the device centerline (upwash), while the device gap region was sim-ultaneously entrained with high momentum fluid from the outer parts of the bounda-ry layer (downwash). This explains the premature and delayed onset of flow separa-tion along the device centerline and device gap regions respectively.

19.5 Concluding Remarks

Numerical and experimental investigations were performed to analyze the influence of Ramped Vanes (RVs) on a flare-induced shock—boundary layer interaction. The studies were performed at Mach 2 over a cone-cylinder-flare model, with a flow turning angle of 24°. An array of RVs was employed ahead of the corner and had a height equivalent to 50% of the local boundary layer thickness.

The counter-rotating vortices emanating from the RVs, produced alternate bands of upwash and downwash regions. The flow separation occurred prematurely in the upwash regions, while the same was delayed along the downwash regions. Furthermore, the vortices penetrated into the interaction region and broke up the separation region into tornado-like vortices. Instantaneous pressure measurements along the downwash region indicated that the separation shock was considerably pushed towards the corner. Alongside this, there was an increase in the separation shock's oscillating frequency range. However, the Strouhal number range was largely preserved.

These observations denote that the RVs are capable of manipulating the shock oscillation frequencies. Also, it has already been demonstrated [22] that the premature onset of separation can be successfully mitigated by increasing the height of these devices.

Acknowledgements This research was supported by the Science and Engineering Research Board, Department of Science and Technology (DST-SERB), through their Core Research Grant Scheme (Grant No: CRG/2018/003925).

References

1. H.A. Cole, A. L. Erickson, A.G Rainey, Buffeting during atmospheric ascent. NASA SP-8001 (1970)
2. L.E. Ericcson, J.P. Reding, Analysis of flow separation effects on the dynamics of a large space booster. J. Spacecrafts Rockets **2**(4), 481–490 (1965)
3. R.D. Blevins, D. Bofilios, I. Holehouse, V.W. Hwa, M.D. Tratt, A.L. Laganelli, P. Pozefsky, P. Pierucci, Thermo-Vibro-Acoustic loads and fatigue of hypersonic flight vehicle structure. AFRL-RB- WP-TR-2009–3139 (2009)
4. D.M. Kuehn, Turbulent boundary layer separation induced by flares on cylinders at zero angle of attack. Technical report R-117 (1961)
5. A. Roshko, G.J. Thomke, Flare induced interaction lengths in supersonic turbulent boundary layers. AIAA J. **14**(7), 873–879 (1976)
6. H. Babinsky, J.A. Edwards, On the incipient separation of a turbulent hypersonic boundary layer. Aeronaut. J. **100**(996), 209–214 (1996)
7. K. Kontis, J.L. Stollery, Incipient separation on flared bodies at hypersonic speeds. Aeronaut. J. **103**(1027), 209–214 (1999)
8. C. L. Running, T. J. Juliano, J. S. Jewell, M. P. Borg, R. L. Kimmel, Hypersonic shock- wave/ boundary-layer interactions on a cone/flare model. AIAA Paper: 2018–3702 (2018)
9. G.T. Coleman, J.L. Stollery, Incipient separation of axially symmetric hypersonic turbulent boundary layers. AIAA J. **12**(1), 119–120 (1973)
10. G.I. Taylor, J.W. Macoll, The air pressure on a cone moving at high speeds—I. Proc. Royal Soc. London Ser. A **139**(838), 278–297 (1933)
11. G.S. Settles, S.M. Bogdonoff, I.E. Vas, Incipient separation of a supersonic turbulent boundary layer at high Reynolds numbers. AIAA J. **14**(1), 50–56 (1976)
12. S. Priebe, M.P. Martin, Low-frequency unsteadiness in shock wave-turbulent boundary layer interaction. J. Fluid Mech. **699**, 1–49 (2012)
13. S.B. Verma, G. Koppenwallner, Unsteady separation in flare-induced hypersonic shock-wave-boundary-layer interaction flowfield. J. Spacecr. Rocket. **39**(3), 467–470 (2002)
14. M.E. Erengil, D.S. Dolling, Effect of sweepback on unsteady separation in Mach 5 compression ramp interactions. AIAA J. **29**(5), 728–735 (1993)
15. S.B. Verma, A. Hadjadj, Supersonic flow control. Shock Waves **25**(5), 443–449 (2015)
16. A.G. Panaras, F.K. Lu, Micro-vortex generators for shock wave/boundary layer interactions. Prog. Aerosp. Sci. **74**, 16–47 (2015)
17. S. Lee, E. Loth, H. Babinsky, Normal shock boundary layer control with various vortex generator geometries. Comput. Fluids **49**, 233–246 (2011)
18. S.B. Verma, C. Manisankar, Assessment of low-profile mechanical vortex generators in controlling a shock induced separation. AIAA J. **55**(7), 2228–2240 (2017)
19. R.R. Martis, A. Misra, Separation attenuation in swept shock wave—boundary layer interactions using different micro vortex generator geometries. Shock Waves **27**(5), 747–760 (2017)
20. S. Lee, E. Loth, On ramped vanes to control normal shock boundary layer interactions. Aeronaut. J. **122**(1256), 1568–1585 (2018)

21. T. Nilavarasan, G.N. Joshi, A. Misra, C. Manisankar, S.B. Verma, Topological modifications due to ramped vanes in a flare-induced shock—boundary layer interaction flowfield. J. Visualization **24**(5), 991–1010 (2021)
22. T. Nilavarasan, G.N. Joshi, A. Misra, C. Manisankar, S.B. Verma, Control of flow separation over an axisymmetric flared body using ramped vanes. Euro. J. Mech.—B/Fluids **95**, 160–177 (2022)
23. T. Nilavarasan, G.N. Joshi, A. Misra, Effect of microramps on flare-induced shock—boundary-layer interaction. Aeronaut. J. **124**(1271), 121–149 (2020)
24. B.H. Anderson, J. Tinapple, L. Surber, Optimal control of shock wave turbulent boundary layer interactions using micro-array actuation. AIAA Paper: 2006–3197 (2006)
25. R. Gejji, I. Walters, S. Beard, A. Lemcherfi, S. Sardeshmukh, S. Heister, C. Slabaugh, Transducer installation effects on pressure measurements in pressure gain combustion devices. AIAA Paper: 2018—0158 (2018)
26. P. Welch, The use of Fast Fourier Transform for estimation of power spectra: A method based on time averaging over short, modified periodograms. IEEE Trans. Audio Electroacoust. **AU-15**(2), 70–73 (1967)
27. K.R. Raman, A study of surface pressure fluctuations in hypersonic turbulent boundary layers. NASA CR-2386 (1974)
28. M.S. Liou, C. Steffen, A New Flux Splitting Scheme. J. Comput. Phys. **107**(1), 23–39 (1993)
29. P.R. Spalart, S.R. Allmaras, A one-equation turbulence model for aerodynamic flows. AIAA Paper: 1992–0439 (1992)
30. J.M. Lighthill, Attachment and separation in three dimensional flow. In: Rosenhead (Ed) Laminar boundary-layer theory, Oxford University Press, Oxford, UK, pp. 77–82 (1963)
31. J.M. Delery, *Three-dimensional separated flow topology: Critical points, separation lines and vortical structures* (John Wiley and Sons Inc., USA, 2013)
32. D.S. Dolling, L. Brusniak, Separation shock motion in fin, cylinder and compression ramp—induced turbulent interactions. AIAA J. **27**(6), 734–742 (1989)

Chapter 20
Numerical Evaluation of the Performance of Supersonic Exhaust Diffuser with and Without Ejected Mass

A. J. Leo, C. V. Sumesh Babu, P. J. Manoj, Nikhil Jacob, L. Rekha, and V. Lijo

Abstract A numerical investigation has been carried out on straight cylindrical supersonic exhaust diffusers (SED) of a high-altitude testing (HAT) to study the nature of vacuum generation with mass ejection from the test cell. Simulations were carried out for SED operating in self-evacuation mode and, SED operating in assisted-evacuation mode (by means of mass ejection). In HAT facilities, the nozzle back-pressure is artificially maintained at a suitably lower level in order to determine the vacuum thrust, maximum heat loads, transient conditions such as engine startup, shut down and re-ignition in vacuum conditions. HAT maintains the required vacuum by arresting the backflow from the atmosphere into the altitude simulation chamber. The steady SED operation is characterized by two distinct operating modes: diffuser unstarting and diffuser starting mode. In the started condition, the vacuum chamber attains its highest vacuum state. Hence, in the case of HAT the most important considerations are the minimum starting and operating pressures. The basic mechanism of SED is the flow of a compressible fluid in a confined duct under the influence of a strong adverse pressure gradient. Wall flow separation under adverse pressure gradient and shocks often results in large recirculation bubbles appearing in SED. Often these recirculation bubbles move towards the low-pressure test cell, preventing the further entrainment of mass from test cell, thereby adversely affecting its vacuum performance. Large improvements in SED performance are possible by controlling or arresting the backflow. The academic community has extensively explored several backflow controlling devices like orifice plate at the inlet of SED, backflow arrester device in vacuum chamber, nozzle lip thickness etc. It is reported that the vacuum ejector start-up is associated with a gradual initial vacuum generation stage, which is followed by a rapid vacuum generation. The slow phase of initial vacuum generation observed in SED can be accelerated by mass ejection. Previous experimental studies with mass-ejection from the vacuum chamber have reported that when mass was

A. J. Leo (✉) · C. V. Sumesh Babu · P. J. Manoj · N. Jacob · L. Rekha · V. Lijo
Department of Mechanical Engineering, Government Engineering College Thrissur, Kerala 680009, India
e-mail: leojose1988@gmail.com

APJ Abdul Kalam Technological University, Kerala, India

© The Author(s), under exclusive license to Springer Nature Singapore Pte Ltd. 2024

G. Rajesh et al. (eds.), *Advances in Shock Interactions*, Springer Proceedings in Physics 408, https://doi.org/10.1007/978-981-97-6099-2_20

ejected from the test-cell, lower pressure was achieved. However, little research has been conducted to show the effect of mass ejection on the starting transients and altitude simulation performance of an ejector.

20.1 Introduction

Supersonic ejector–diffuser systems are often used to create very high vacuum levels in the vacuum/test chamber in which the rocket engine is placed and the ejector is self-pumping with hot exhaust gas from the rocket nozzle.

In its simplest form, HAT consists of a straight cylindrical duct attached to a (vacuum) chamber surrounding the rocket engine. The momentum of the engine exhaust is utilized to lower the nozzle back-pressure. Since no other mechanical pumping devices are used to create vacuum, this is called self-evacuation mode. Static pressure recovery in the diffuser is accomplished through momentum exchange caused by a shock-wave system and deceleration to subsonic velocity.

HAT maintains the required vacuum by arresting the backflow from the atmosphere into the altitude simulation chamber. The HAT operation is characterized by two distinct operating modes: diffuser un-starting and diffuser starting mode. Under the started condition, the exhaust plu me from the nozzle impinges on the duct wall and the complex system of branched shocks fully established serves to isolate the vacuum chamber from the downstream ambient state. In the un-started mode, the shock train does not fill the entire cross section of the duct, and consequently, the vacuum chamber is not shielded from the ambient conditions. In the started condition, the vacuum chamber attains its highest vacuum state. Hence, in the case of HAT the most important consideration is the minimum starting pressure. Wall flow separation under adverse pressure gradient and presence of strong shocks often results in large recirculation bubbles appearing in SED. Often these recirculation bubbles move towards the low-pressure test cell, preventing the further entrainment of mass from test cell into the supersonic primary-jet, thereby adversely affecting its vacuum performance. Large improvements in SED performance are possible by controlling or arresting the backflow. The academic community has extensively explored several backflow controlling devices like orifice plate at the inlet of SED, backflow arrester device in vacuum chamber, nozzle lip thickness etc [1].

The HAT facility, is basically designed to test a particular rocket motor, hence the level of vacuum that can be achieved remains invariant with fixed geometry of HAT and operating conditions. Present work concentrates on HAT performance improvement which allows rocket engines to be tested at much higher simulated-altitude (vacuum) conditions without altering the basic geometry of the HAT.

It is reported that the vacuum ejector start-up is associated with a gradual initial vacuum generation stage, which is followed by a rapid vacuum generation [1]. The slow phase of initial vacuum generation observed in SED can be accelerated by mass ejection. Typical mass-ejection from the vacuum chamber is shown in Fig. 20.1. J W Hale [2] has done experimental investigation of ejector-diffuser with and without

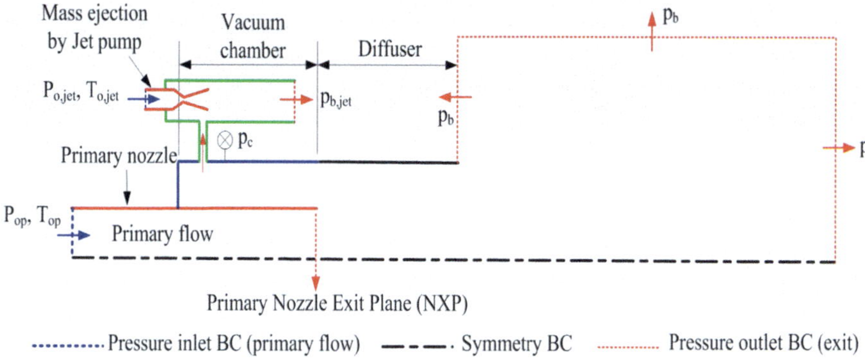

Fig. 20.1 Schematic of HAT with mass-ejection from vacuum chamber with the help of two jet pumps

mass ejection and reported that when mass was ejected from the test-cell, lower pressure was achieved. However, little research has been conducted to show the effect of mass ejection on the starting transients and altitude simulation performance of an ejector. The baseline geometry of the SED is adapted from the experimental work of Arun Kumar and Rajesh [3] and mass-ejection is implemented in this constant-area mixing ejector.

In the started mode, owing to such shielding effect of supersonic jet from downstream conditions, mass ejection from the test cell will be effective in lowering test cell pressure. Hence, the external jet pumps play supporting role only; it is effective in the started-mode only. Hence in the un-started mode both the jet pumps could even be switched off. Present study reports only the steady state operation of HAT with jet pumps.

In the transient operation of ejector [3] it is reported that splitting up of recirculation bubble near the primary-nozzle exit caused a rapid decrease in test-cell pressure. Unsteady operation of HAT as primary pressure ramps up from ambient pressure to high pressure, with mass ejection at multiple points at the inlet of diffuser can cause splitting-up of the recirculation bubble, causing performance enhancement. In the present work, shock structure and complex wall-bounded flows in HAT with and without mass ejection is investigated for steady state operation in detail. Results from the transient operation of HAT will be reported later with particular attention to the development and splitting-up of recirculation flows.

20.2 Computational Details

The basic dimensions of ejector (Fig. 20.1) are taken from a study on two-dimensional supersonic ejector by Arun Kumar, and Rajesh [3] Two-dimensional, steady numerical simulations were performed using the commercial CFD software ANSYS

Fluent16. The working substance used in the present study is dry air with ideal gas assumption. The *k*–*ω* SST is taken for the turbulence model and all the walls are considered to be adiabatic with no-slip.

The primary stagnation pressure (P_{0p}) is varied from 2.5 bar to 5.0 bar. From the experiment [2] HAT starting pressure is 4.5 bar. The backpressure, p_b is fixed at 1.0 bar as shown in Fig. 20.1. Stagnation temperature is taken as 300 K for the primary stream. As a two-dimensional supersonic ejector is considered for computational analysis, only half a portion is taken for discretization due to symmetry. The boundary conditions are also shown in Fig. 20.1. Several mass ejection strategies (from vacuum chamber or diffuser inlet) by ejecting different amount of masses were explored and it is found that mass ejection of about 3.0% of rocket-engine flow from the test-cell can result in significant reduction in test-cell pressure.

The grid system consisted of structured quadrilateral cells (Fig. 20.2). Three sets of grids are used to determine if there are sufficient grid points. The coarse grid has about 50,000 cells. The medium grid set has 110,000 cells The fine grid has 246,000 cells. The results of grid refinement study (not shown here) showed no substantial differences with regard to the evolution of centerline Mach number with medium and fine grids. Hence, the results presented in this work are from the medium grid set.

Shown for comparison are the primary pressure (Fig. 20.3) predicted with the code and experiment [2]. Reasonable agreements were found between CFD and experiment. Starting pressure is under predicted by code, suggesting that wall effects are predominant. Better match with the experiment may be achieved by a full-scale 3D simulation. However, the agreement between the predicted chamber-pressure (Pc) and the experimental values provides reasonable trend, and this validated model has been used to obtain flow features.

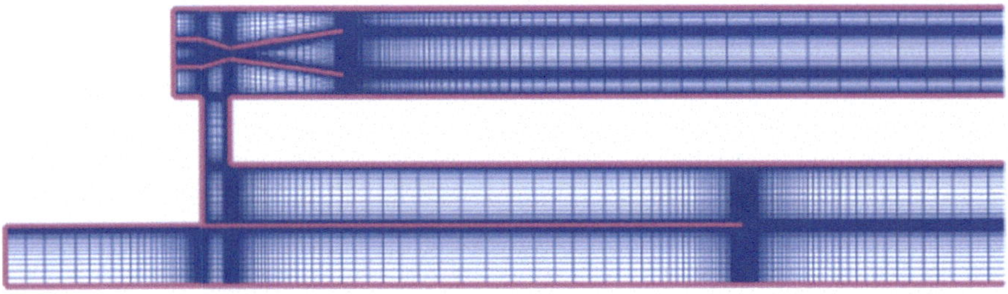

Fig. 20.2 Computational mesh

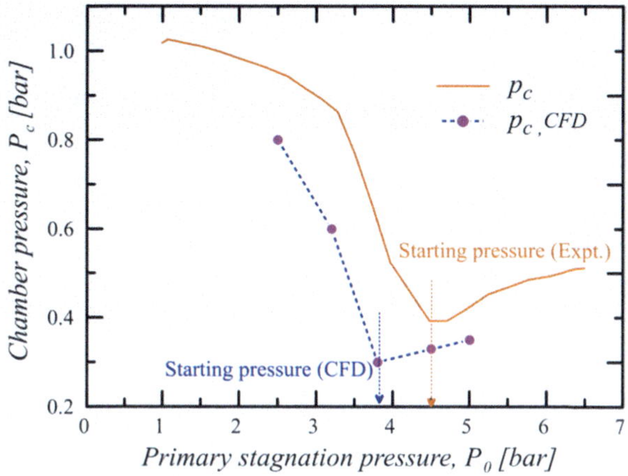

Fig. 20.3 Numerical validation with the experimental results

20.3 Results and Discussion

Figure 20.4 shows the development of the flow as primary pressure (P_{0p}) is kept constant at 5 bar, for which the HAT is in the started condition, but the jet pump operating pressure is changed. Two cases were investigated jet-pump total pressure is 6 bar and 8 bar. With the two jet pumps off, the vacuum pressure Pc = 64.6 kPa. With jet pumps operating with 6 bar supply pressure, the vacuum is improved to 41 kPa. Upon further increasing the jet-pump pressure the vacuum achieved is 34.5 kPa. Hence, without modifying the geometry of HAT, the same facility can simulate higher altitudes, with the addition of two supplementary jet-pumps. For the lowest vacuum achieved, the two jet-pumps together extract a mass of 3% of the rocket nozzle flow.

The shock structure with and without mass ejection is investigated in detail in Fig. 20.5. The numerical schlieren (gradient of density along streamwise direction). The darker regions represent possibility of shock, while lighter shade may indicate expansion waves. As Pc is reduced with increasing jet-pump driving pressure, the primary-jet becomes more and more under expanded, resulting in stronger shock induced boundary layer separation at the foot of the oblique shock (Fig. 20.5c). as the primary-jet impingement angle increases, the entire shock reflection pattern changes. Two shock cells can be seen for no ejected mass case, whereas with mass ejection three shock-cells are seen, which increase the static pressure recovery of the diffuser [4–8].

To numerically visualize the recirculation patterns, stream lines (originating from primary inlet and vacuum chamber) is plotted in Fig. 6. As shock strength increases, the recirculation bubble present near the jet impingement region also enlarges. This recirculation bubble can be split in to two if we move the jet-pumps towards the NXP. Altering the shape of the recirculation bubble can change the evacuation

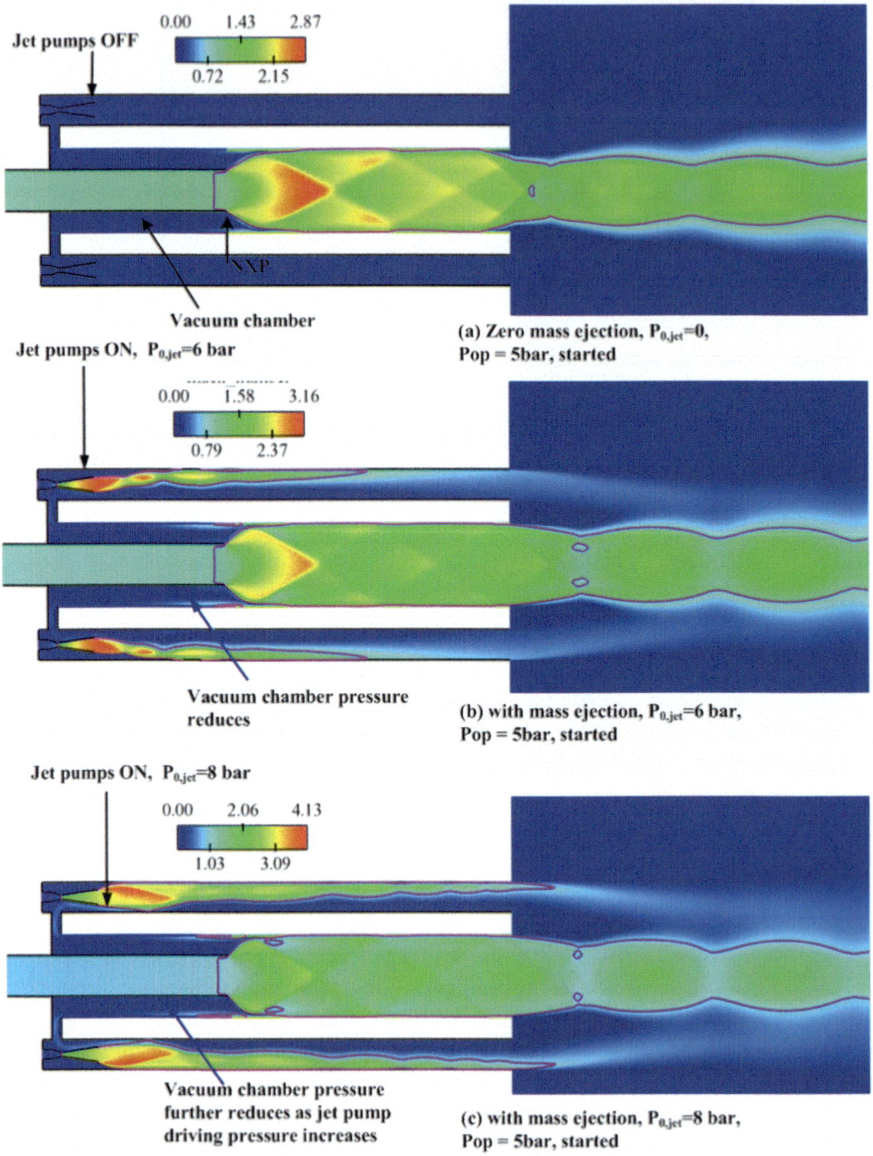

Fig. 20.4 Mach contours with overlaid sonic lines (pink) for jet pumps ON/OFF for a fixed primary pressure , $P_{0p} = 5bar$

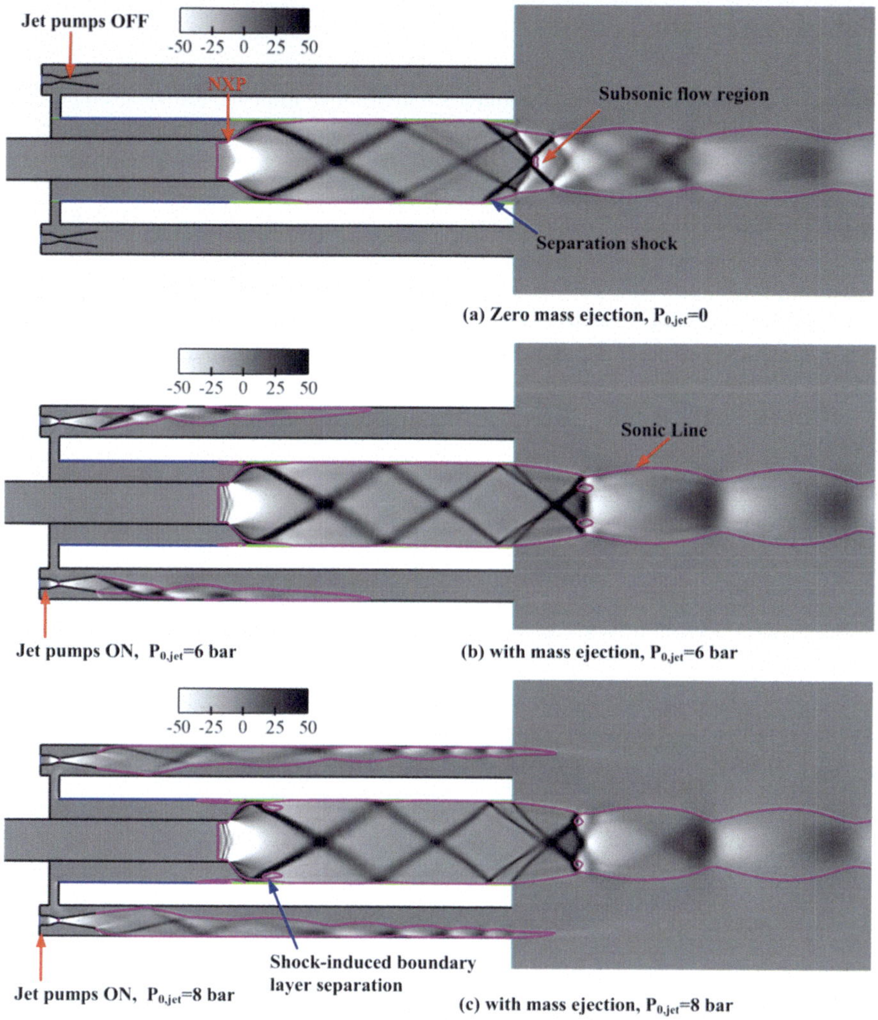

Fig. 20.5 Numerical schlieren with and without mass ejection for a fixed primary pressure , $P_{0p} = 5bar$

performance and hence the level of vacuum attained. Hence, the location of mass ejection from the vacuum chamber is crucial. The location of mass ejection will be optimized in future studies.

20.4 Conclusion

The present work numerically investigates a method to improve the vacuum performance of an HAT by mass ejection from test cell. Numerical simulations are carried out using a 2-*D* supersonic ejector employing shear stress transport k-ω model with dry air as the working substance. Since the present model has low aspect ratio, better match with the experiment may be achieved by a full-scale 3D simulation. For the operating pressure ranges studied, the maximum enhancement resulted with 3.0% ejection of total mass flow rate of the rocket engine. The flow development is visualized by Mach contours with and without mass ejection. Numerical schlieren and stream line plots help to understand the shock and recirculation patterns. Since mass ejection is essentially an unsteady process, unsteady simulations with mass ejection at single or multiple points will be reported in future studies.

References

1. V. Lijo, H. Dong, S. Matsuo, T. Setoguchi, A study of the supersonic ejector—diffuser system with an inlet orifice. Aerosp. Sci. Technol. **23**(1), 321–329 (2012). https://doi.org/10.1016/j.ast.2011.08.009
2. J.W. Hale, Ejector-diffuser performance with by, **40**(600), (1964)
3. R. Arun Kumar, G. Rajesh, Physics of vacuum generation in zero-secondary flow ejectors. Phys. Fluids (2018). https://doi.org/10.1063/1.5030073
4. H.W. Emmos, Fundaments of Gas Dynamics, Vol. II of Series, High Speed Aero dynamics and Jet Propulsion, Princeton University Press, Princeton, NJ, (1958)
5. A.H. Shapiro, The Dynamics and Thermodynamics of Compressible fluid flow, Vol I, Ronald Press, New York, (1953).
6. J. Fabri, J. Paulon, Theory and Experiments on Supersonic Air-to-Air Ejectors, NACA TM 1410, (1958)
7. P.J.Waltrup, F. S.Biling, Structure of shock waves in cylindrical ducts, AIAA Journal **11**(10), 1404–1408 (1973)
8. Annamalai K, Visvanathan K, Sriramulu V and Bhaskaran K A, "Evaluation of the Performance of the Supersonic Exhaust Diffuser Using Scaled down models", *Experimental Thermal and Fluid Science*, **17**(3), 217–229 (1998)

Chapter 21
Detonation Dampers for Ducts Transporting Gaseous Fuels

Zbigniew A. Walenta and Agnieszka M. Slowicka

Abstract The very serious problem connected with long distance transport of gaseous fuels is connected with the fact that detonation may occur inside the duct if some air leaks into it. Detonation is particularly dangerous for compressors which "push" the gas through pipelines. It is therefore necessary to use some "detonation dampers" to protect these compressors. The commonly used detonation damper has a form of a matrix of narrow channels, placed across the pipe transporting gas. Detonation wave is supposed to be extinguished due to cooling by cold walls of these channels. To achieve efficient damping the channels should be very narrow. Our earlier simulations (Walenta et al. in Central European Journal of Energetic Materials. 1:49–59, 2004; Walenta Z.A., Slowicka A.M.: Extinguishing detonation in pipelines—optimization of the process, In: K. Kontis (ed.), Proceedings of 22nd Int. Shock Interaction Symposium, Glasgow, pp. 202–206. (2016).; Walenta Z.A., Slowicka A.M.: Detonation Waves in Narrow Channels of Various Shapes, Proceedings of 23rd Int. Shock Interaction Symposium, Skukuza, Kruger National Park, South Africa, pp. 135–140 (2018).) indicate, that preferable widths should be of the order of 0.005 mm. This is not acceptable for practical reasons—in real applications it is close to 0.5 mm. In such channels cooling the gas by heat transfer is inefficient and cannot extinguish the flame (cooling effect is proportional to perimeter of the channel cross-section and amount of gas to be cooled to cross-section surface). Rarefaction wave generated behind sharp increase of channel cross-section is an alternative phenomenon, which may be used to cool the burning gas (Walenta et al. in Central European Journal of Energetic Materials. 1:49–59, 2004). In our earlier papers we tested several channel shapes with increase of cross-section (Walenta Z.A., Slowicka A.M.: Extinguishing detonation in pipelines—optimization of the process, In: K. Kontis (ed.), Proceedings of 22nd Int. Shock Interaction Symposium, Glasgow, pp. 202–206. (2016).; Walenta Z.A., Slowicka A.M.: Detonation Waves in Narrow Channels of Various Shapes, Proceedings of 23rd Int. Shock Interaction Symposium, Skukuza, Kruger National Park, South Africa, pp. 135–140 (2018).). Here we

Z. A. Walenta (✉) · A. M. Slowicka
Institute of Fundamental Technological Research, Polish Academy of Sciences, Pawinskiego 5B, 02-106 Warszawa, Poland
e-mail: zwalenta@ippt.pan.pl

© The Author(s), under exclusive license to Springer Nature Singapore Pte Ltd. 2024

G. Rajesh et al. (eds.), *Advances in Shock Interactions*, Springer Proceedings in Physics 408, https://doi.org/10.1007/978-981-97-6099-2_21

compare channels 0.5 mm wide, without and with sharp increases and decreases of cross-section. Such channels were found to be the best for practical applications.

21.1 Introduction

One of the serious problems connected with long distance transport of gaseous fuels is connected with the fact that detonation may occur inside the duct if some air leaks into it. Detonation is particularly dangerous for compressors which "push" the gas through pipelines. It is therefore necessary to use some "detonation dampers" to protect these compressors.

The commonly used detonation damper has a form of a matrix of narrow channels, placed across the pipe transporting gas. Detonation wave is supposed to be extinguished by cooling with cold walls of these channels. To achieve good efficiency of damping the channels should be very narrow.

Our earlier simulations [1–3] indicate, that the width of the damper channels, preferable from the point of view of damping efficiency, should be of the order of about 0.005 mm. Such width is, unfortunately, not acceptable for practical reasons. The width of the channels of real dampers is usually close to about 0.5 mm. In such channels cooling the gas by heat transfer to the walls is relatively inefficient and cannot extinguish the flame (cooling effect is proportional to perimeter of the channel cross-section, the amount of gas to be cooled is proportional to the cross-section surface, proportional to square of perimeter).

Rarefaction wave generated behind sharp increase of the channel cross-section is an alternative phenomenon, which may be used to cool the burning gas [1]. In our earlier papers we tested several shapes of channels with increase of cross-section [2, 3]. We found, that channels of the shape shown in Fig. 21.1—bottom were the most favorable. Here we compare results for channels 0.5 mm wide, without and with sharp increases and decreases of cross-section (Fig. 21.1). Channels of such dimensions are most interesting for practical application.

There is one point, which should be noted here. The device for extinguishing detonation has a form of a thick steel plate (of the order of 50 mm thickness), with large number of holes of diameter equal to a fraction of a millimeter. It is impossible to drill large number of such holes in such plate. Instead, such device is usually manufactured from thin steel sheets, with narrow and shallow grooves, pressed firmly together. Such method of manufacturing makes it possible to produce channels of nearly arbitrary size and shape.

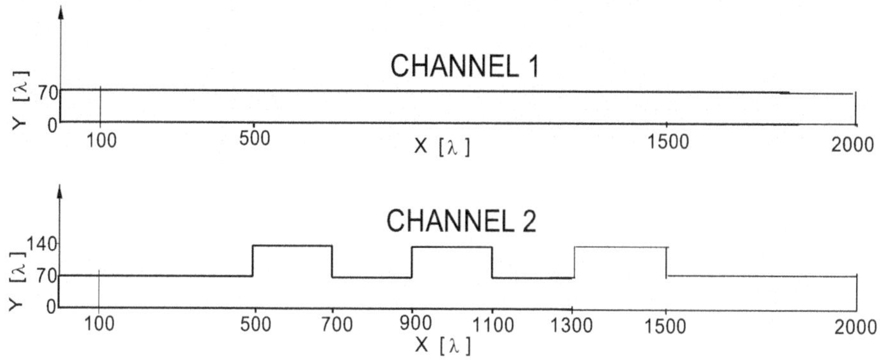

Fig. 21.1 Shapes of the considered channels. (λ—mean free path at initial conditions)

21.2 Method of Simulation

The simulations presented in this paper were performed with the standard Direct Simulation Monte Carlo (DSMC) technique [4]. The DSMC technique makes it possible to simulate flows in various geometrical configurations and it also offers a possibility of taking into account relaxation phenomena and chemical reactions [4, 5].

This, unfortunately, increases complexity of the computer programs and the necessary computing times. However, in the considered case considerable simplifications can be made because in a detonation wave the medium is far from thermodynamic equilibrium, combustion proceeds very fast and relaxation processes at the molecular level may be disregarded. The factor of prime importance is the produced thermal energy.

21.3 Model of a Detonating Medium

In the present work we investigate only the influence of shape and size of the channel upon the flow inside it, and upon the process of extinguishing detonation. A very simple model of the detonating medium [1, 6] may therefore be used. All molecules of this medium are identical, hard spheres. Part of them carry certain amount of "internal" energy, the remaining are "inert"—carry no "internal" energy and cannot accept it in any way. The "internal" energy of a molecule may be transformed into kinetic energy during collision with an "inert" molecule, if the colliding molecules approach each other with high enough "velocity of approach" (Fig. 21.2). The relative velocity of the molecules is then increased suitably.

The "internal" energy of a single molecule was such, that the relative velocity of colliding molecules was increased by the value equal to 10 times the most probable molecular speed.

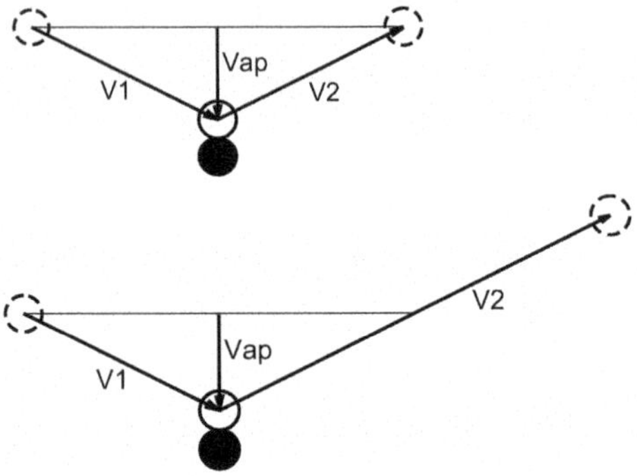

Fig. 21.2 Collisions of two molecules in reference frame connected to one of them. Top—elastic, bottom—with energy release; V1—relative velocity before collision, V2—relative velocity after collision, Vap—"velocity of approach"

21.4 Interactions with Solid Walls

Interactions of the molecules with walls were simulated with the simple model introduced by J.C. Maxwell [7]: molecules reflect from the walls either specularly (without exchange of tangential momentum and energy) or diffusely (molecules are adsorbed by the wall and re-emitted in directions selected at random, with energies corresponding to temperature of the wall). The "accommodation coefficient" α, (ratio of the number of molecules reflected diffusely to total number of reflected molecules) may vary from 0 to 1. The value $\alpha = 0$ (specular reflections) corresponds to no exchange of tangential momentum and energy, the value $\alpha = 1$ (diffuse reflections)—to maximum possible exchange (maximum friction and cooling). For majority of the so-called "technological surfaces" the accommodation coefficient is close to unity.

21.5 Simulation of Flow in Channels of Larger Cross-Section

According to the Maxwell's model of interactions, the walls influence the flow through diffuse reflections of the molecules only. The influence of walls upon the flow in channel depends therefore only on the ratio of the number of molecules reflecting diffusely to total number of molecules in the flow. Changing the number of molecules reflected diffusely (i.e. the assumed accommodation coefficient) it is possible to obtain, in the narrow channel, the flow picture similar to that in the channel of larger cross-section. The required value of the accommodation coefficient is then inversely proportional to hydraulic radius of the simulated channel.

The above does not apply to the narrowest channels, where the flow is close to free-molecular. The criterion is the Knudsen number, which should not be larger than 0.01–0.02.

21.6 Details of Simulations

We present the results of simulations of behavior of the stationary flow and detonation waves in two channels:

Channel 1 (Fig. 21.1—top)—of size 70 × 70 units and 2000 units long (the unit of length equal to 1 mean free path of the molecular motion at initial conditions).

Channel 2 (Fig. 21.1—bottom)—also 70 units wide; other dimensions given in Fig. 21.1.

The interior of the channels was divided into cubic cells of dimension equal to 1 unit. Each cell contained initially about 5 simulated molecules.

In each channel, 100 units from its left end, there was a "diaphragm" separating the hot driving gas from the rest of the channel. Temperature of the driving gas was 22.5 times higher than temperature of the gas in front of the "diaphragm" and the number density was equal to about 0.9 of the number density in that region. Removal of the "diaphragm" produced a shock wave strong enough to be transformed sufficiently quickly into a detonation wave.

500 units from both ends of each channel the molecules were reflected from the walls specularly. Such region of flow without losses behind the "diaphragm" was necessary for the detonation to develop. At the other end of the channel it was assumed to enable checking whether the extinguished detonation would not reappear.

In the central part of each channel the molecules were reflected from the walls with accommodation coefficient $\alpha = 0.01$. At the dimension of the channel cross-section equal to 70 units, i.e. ~ 5 μm, this corresponds to simulated channels acceptable technologically (dimension of the cross-section ~ 0.5 mm).

Simulation consisted of two stages. During the first stage—generation of the stationary flow—the molecules were allowed to escape freely from the low-pressure end of the channel. The escaping molecules were then transferred to the place at low-pressure side of the "diaphragm", which created certain pressure difference between the ends of the channel and, after sufficiently long time (usually not less than about 150 h of computing time) it led to establishing stationary flow.

Second stage of simulation was initiated by the "diaphragm" removal. It produced the shock wave, which was then transformed into a detonation wave.

21.7 Results

21.7.1 Stationary Flow

Figure 21.3 shows distributions of flow parameters along both considered channels, at accommodation coefficient $\alpha = 0.01$ in stationary flow. Figure 21.3—upper, left, shows the distributions of density, Fig. 21.3—upper, right—velocity, Fig. 21.3—lower, left—pressure, and Fig. 21.3—lower, right—flow rate (product of local density and velocity). In all these figures density and pressure are related to their initial values (before beginning of the flow) and velocity—to the initial value of the most probable velocity of the molecular motion. All parameters are averaged over the local cross-section of the channel.

The presented variations of density, velocity and pressure along the channel in stationary flow are appreciable. The pressure drop along the straight channel 1 is smaller than that along channel 2 (Fig. 21.3, lower, left) and the flow rate is larger (Fig. 21.3, lower, right). This means, that resistance to the flow in the straight channel is smaller than that in the channel 2 (which was to be expected).

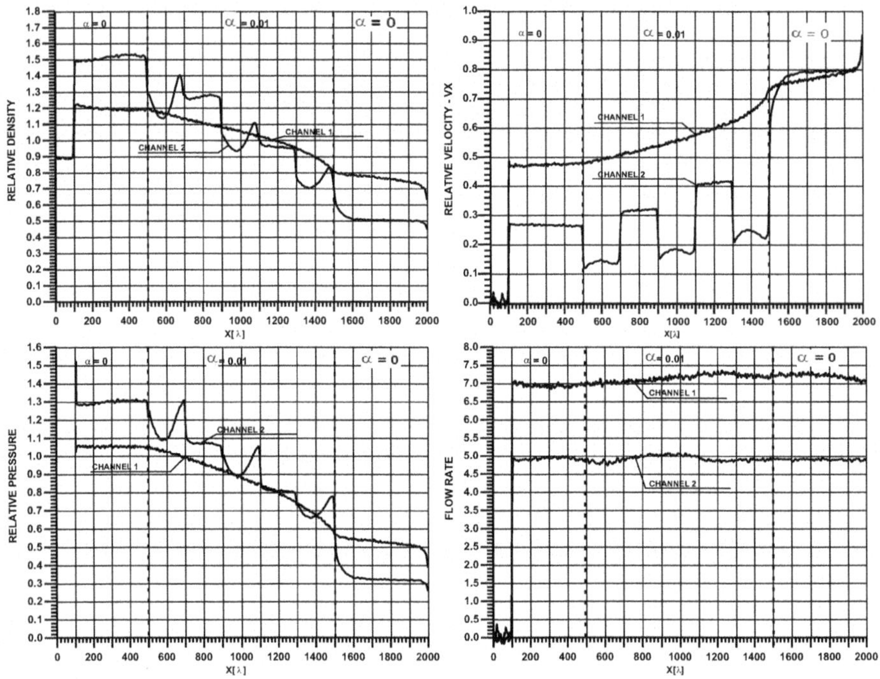

Fig. 21.3 Stationary flow parameters in considered channels; accommodation coefficient $\alpha = 0.01$. Upper, left—density, upper, right—velocity, lower, left—pressure, lower, right—flow rate

To make it more precise, the coefficients of hydraulic resistance were calculated from the standard formula:

$\zeta = 2\Delta p/(\rho u^2)$.

where Δp—pressure drop along the channel, ρ—density and u—velocity in region $100 < x < 500$. The calculated coefficient of hydraulic resistance of channel 2 is about 5 times larger than that of the straight channel 1.

21.7.2 Detonation

In this section we present diagrams, uniformly distributed in time, of temperature and percentage of flammable component along both considered channels, resulting from simulation of the behavior of detonation wave in quiescent medium and in stationary flow inside the considered channels. The amount of "energetic" molecules in all channels is equal to 7%.

In Fig. 21.4—left, the diagrams of temperature inside channel 1 at $\alpha = 0.01$ and no flow in front of detonation wave are shown. The detonation wave, fully formed in the initial part of the channel after "diaphragm" removal, in its central part is affected very weakly—its amplitude decreases by about 9% only. In the end part of channel, where no friction and heat exchange are present, it begins regaining its intensity.

In Fig. 21.4—right the diagrams of temperature inside channel 1, under conditions of stationary flow in front of detonation wave are shown. As before, the detonation is fully formed after "diaphragm" removal. In the central part of the channel the

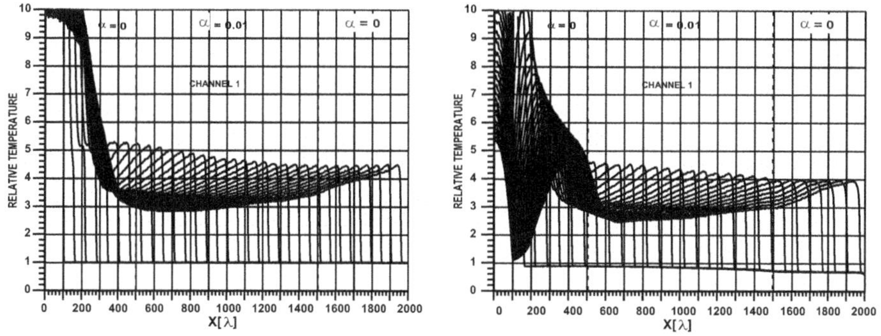

Fig. 21.4 Temperature distributions in channel 1 at $\alpha = 0.01$. Medium contains 7% of "energetic" molecules. Left: detonation in quiescent medium. Right: detonation in stationary flow

maxima of temperature decrease by about 30%. Still, the ratios of this temperature to the values in front of detonation wave increase, which means that intensity of detonation actually grows. This is result of density decrease along the channel in stationary flow (compare Fig. 21.3 upper, left).

In Fig. 21.5—upper, left the diagrams of temperature inside channel 2 in progressing detonation wave at no flow conditions are shown. The detonation wave is fully formed after "diaphragm" removal and in the central part of the channel (where $\alpha = 0.01$ and cross-section varies) its amplitude gradually decreases. Close to the end of channel (in region where $\alpha = 0$) it transforms into weak shock wave followed by some region of increasing temperature. Unfortunately, the strength of this shock grows with time.

Information on what actually happens there can be found in Fig. 21.5, upper, right. This picture shows distributions of percentage of flammable component of the medium, corresponding to temperature distributions in Fig. 21.5, upper, left. Close to the end of channel (in region where $\alpha = 0$) the shapes of the curves suggest, that main part of the medium is not burning, however flame is still there and the burning area grows.

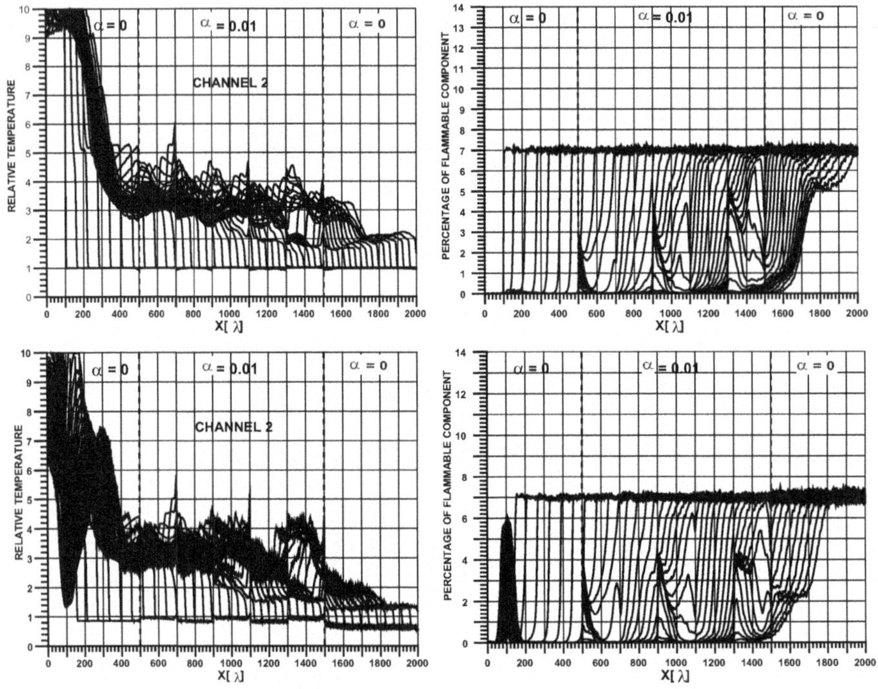

Fig. 21.5 Distributions of temperature (left) and percentages of flammable component (right) in channel 2, at $\alpha = 0.01$. Medium contains 7% of "energetic" molecules. Top: detonation in quiescent medium. Bottom: detonation in stationary flow

In Fig. 21.5—lower, left the diagrams of temperature inside channel 2, in detonation wave progressing into area of stationary flow are shown. Formation of the detonation and its behavior in the central part of the channel are here similar to those in the previous case. However, in the end section of the channel, where $\alpha = 0$, the strength of the emerging weak shock wave is nearly constant and its distance to the following area of temperature increase is growing with time. Figure 21.5, lower, right (similar to Fig. 21.5, upper, right) does not show the flame there.

21.8 Conclusions

- The proposed method of DSMC simulation of flow in channels of a detonation damper was found to work also for channels of technologically acceptable dimensions.
- Presence of abrupt changes of cross-section of narrow channels used in detonation dampers generally increases efficiency of damping. In channels of technologically acceptable dimensions this influence seems to be much stronger than influence of heat exchange with walls.
- In channels of technologically acceptable dimensions (~ 0.5 mm width, or more) the increase of flow resistance due to changes of cross-section is significant, however, taking into account the increased efficiency of damping, the net gain seems worthwhile.

References

1. Z.A. Walenta, A. Teodorczyk, A. Dabkowski, W. Witkowski, Direct Monte—Carlo simulation of a detonation wave in a narrow channel, containing flammable gas. Central Eur. J. Energ. Mater **1**(1), 49–59 (2004)
2. Walenta Z.A., Slowicka A.M.: Extinguishing detonation in pipelines—optimization of the process, In: K. Kontis (ed.), Proceedings of 22nd Int. Shock Interaction Symposium, Glasgow, pp. 202–206. (2016)
3. Walenta Z.A., Slowicka A.M.: Detonation waves in narrow channels of various shapes, Proceedings of 23rd Int. Shock Interaction Symposium, Skukuza, Kruger National Park, South Africa, pp. 135–140 (2018)
4. G.A. Bird, *Molecular gas dynamics and the direct simulation of gas flows* (Clarendon Press, Oxford, 1994)
5. Larsen P.S.: Borgnakke C., Paper A7 in M. Becker and M. Fiebig (ed.), Rarefied Gas Dynamics 1. DFVLR Press, Porz–Wahn, Germany (1974)
6. Z.A. Walenta, K. Lener, Direct Monte—Carlo simulation of developing detonation in gas. Shock Waves **18**, 71–75 (2008)
7. Maxwell J.C., In W.D. Niven, (ed.), The scientific papers of James Clerk Maxwell 2, 706. New York, Dover (1952)
8. Walenta Z.A., Slowicka A.M., Optimization of detonation dampers for ducts transporting gaseous fuels. Proceedings of 32nd International Symposium on Shock Waves, Singapore, pp. 547–555 (2019)

Chapter 22
Preliminary Studies on Continuously Varying Mach Number Tunnel

V. Vijayakrishnan, Madhura Narayana Sabhahit, and T. M. Muruganandam

Abstract In this paper, shock-shock interactions are investigated experimentally, in a varying Mach number flow. The tunnel was operated in accelerating and decelerating modes, with average dM/dt of 1.67. The shock angles were used to obtain the Mach number upstream from the steady gas dynamics expressions, and compared with pressure ratio based Mach number at the exit. It was found that the Mach numbers thus obtained matched well, within the error bars, for both accelerating and decelerating cases. This suggests that the dM/dt of 1.67 is not sufficient to invoke unsteady gas dynamics behavior. This was also corroborated by straight line shockwaves.

Keywords Shock-shock interaction · Variable mach number tunnel · Accelerating flow · Decelerating flow

22.1 Introduction

Shock wave interactions occur in many physical applications in high-speed flows. Shock wave interactions are observed in supersonic inlets, isolators, and nozzle flows. These interactions have a significant effect on the performance of these devices. In steady flows, two interactions are possible, (1) regular interaction (RI) and (2) Mach interaction (MI), as shown in Fig. 22.1. As the vehicle accelerates or decelerates, any one of the interactions can occur. The interaction is a function of Mach number and flow deflection angle. During acceleration or deceleration of the flight, the vehicle Mach number changes and for fixed flow deflection angle, RI↔MI transition occurs. In the shock wave research community, this phenomenon is referred to as Mach number-based hysteresis [1].

V. Vijayakrishnan (✉) · M. N. Sabhahit · T. M. Muruganandam
Department of Aerospace Engineering, IIT Madras, Chennai, India
e-mail: vjk.surya@gmail.com

© The Author(s), under exclusive license to Springer Nature Singapore Pte Ltd. 2024

G. Rajesh et al. (eds.), *Advances in Shock Interactions*, Springer Proceedings in Physics 408, https://doi.org/10.1007/978-981-97-6099-2_22

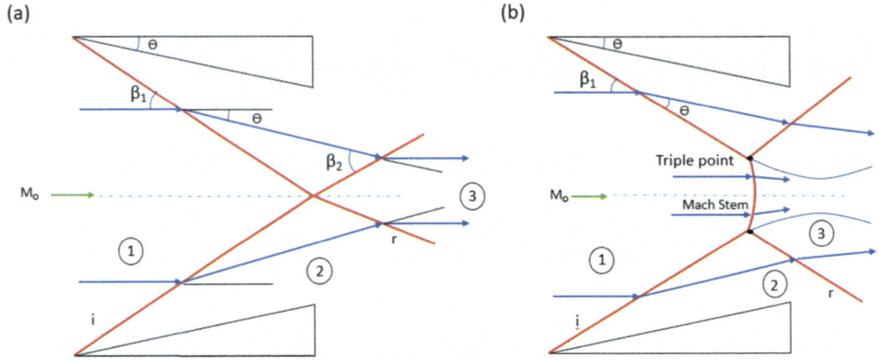

Fig. 22.1 Schematic of **a** regular interaction, **b** Mach interaction

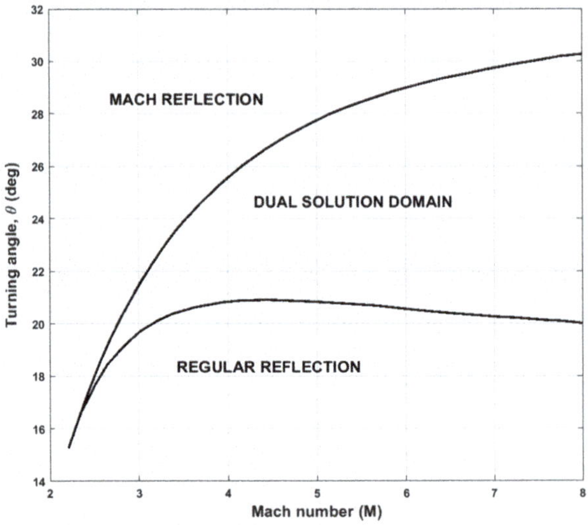

Fig. 22.2 Plot of von Neumann and detachment criteria for various wedge angles and Mach number

The von Neumann and detachment criteria set boundaries for the RI↔MI transition, as shown in Fig. 22.2. Only RI is possible in the region below the von Neumann line, and in the region above the detachment line, only MI is possible. The area between the von Neumann and the detachment line is the dual-solution domain; both RI and MI are possible.

Li and Ben-Dor [2] solved the interaction between an oblique shock wave and a centered expansion wave in steady flows analytically. Ivanov et al. [3] investigated Mach number variation-induced hysteresis numerically and reported both RI↔MI and MI↔RI depending on the direction of Mach number. Durand et al. [4] investigated shock-shock interactions between two wedges experimentally and numerically. They reported that RI→MI and MI→RI take place in the middle of the dual-solution domain. Yao et al. [5] studied supersonic and hypersonic intake flow under off-design conditions. They showed that shock-expansion interactions signif-

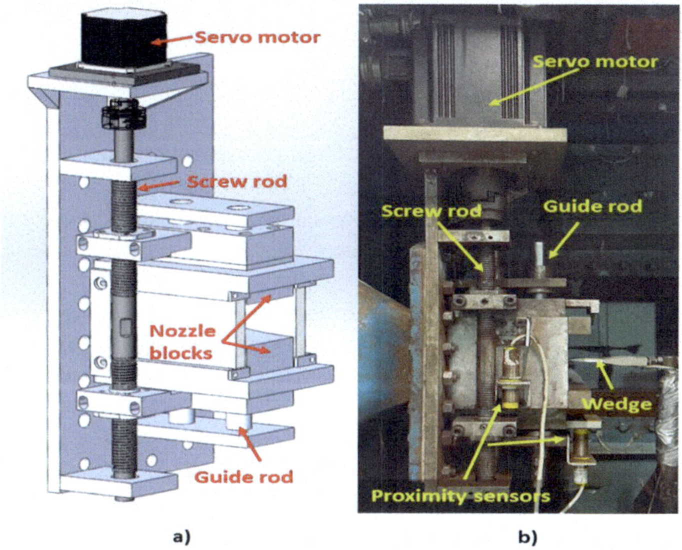

Fig. 22.3 Translation nozzle assembly **a** CAD model, **b** Actual nozzle assembly

icantly alter the dual-solution domain. They also showed that the Mach stem height is an increasing function of the turning angle. Nel et al. [6] studied the interaction between an expansion fan and a shock wave with various schlieren techniques. They used three cut-off techniques to check the interaction's various flow features, and each method's efficacy is reported. Goyal et al. [7] studied the dynamic effects of RR→MR in steady flow computationally. They investigated the unsteady changes in the Mach stem height, position of the reflection point, and shock angle at the triple point on the wedge for various rotation rates. They reported that the pivot point and rotation rate significantly affect the Mach stem growth and the movement of the reflection point at higher Mach numbers. Laguarda et al. [8] presented a post-processing methodology for accurate tracking of the shock transition between RI↔MI. They also reported the three-dimensional side wall effects with variable focal plane focusing schlieren system. The present paper investigates shock-shock interactions in the variable Mach flow. The variation in the shock angle changes with varying Mach numbers has been studied using time-resolved schlieren imaging and pressure measurements. The present paper is arranged into three parts. The first part discusses the experimental facility and techniques used in this study. The second part reports the result and discussion section, in which the shock angle changes in accelerating and decelerating flows have been discussed. The conclusions drawn from the preliminary studies on the variable Mach number flow are reported in the end section (Fig. 22.3).

22.2 Experimental Setup

The Continuously Varying Mach Number Tunnel (CVMT) at the National Centre for Combustion Research and Development (NCCRD), Indian Institute of Technology Madras, is used for the present study [9]. High-pressure air is supplied to the wind tunnel from the compressor facility available at the NCCRD. The nozzle's area ratio sets the flow's acceleration or deceleration inside the wind tunnel during the run. A range of continuously varying Mach numbers from 2 to 2.79 is attainable in the wind tunnel.

In our present study, two models were used to study shock-shock and shock-expansion interactions. The width of the models is 40 mm. The leading edge of the shock wave generators is 14 mm apart from each other. The shock wave generators are symmetric with a turning angle of 10° as shown in Fig. 22.4.

In the shock-expansion interaction model, the expansion fan generator and the shock wave generator are designed with a turning angle of 10° as shown in Fig. 22.5.

The flow field is visualized using time-resolved high-speed schlieren imaging with Photron FASTCAM SA-Z camera. Settling chamber stagnation pressure and nozzle exit wall static pressure was acquired using Scanivalve DSA 3271 (0–145 psid) with a sampling frequency of 500 Hz. The exit pressure in the nozzle was measured by a port at 2 mm upstream of the exit plane.

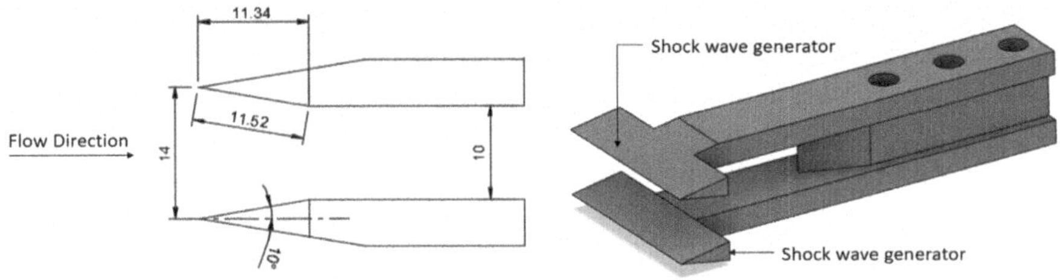

Fig. 22.4 Schematic and the three-dimensional model for the shock-shock interaction

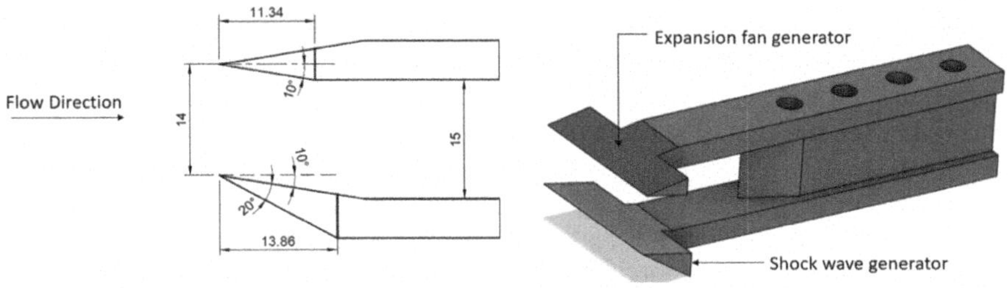

Fig. 22.5 Schematic and the three-dimensional model for the shock-expansion interaction

22.3 Results and Discussion

Shock-shock and shock-expansion interactions are investigated using the model geometry shown in Figs. 22.4 and 22.5. The instantaneous images of the schlieren for acceleration and deceleration for shock-shock interaction are shown in Figs. 22.6 and 22.7. The constant length red line is marked in the schlieren images to better visualize the shock impingement point moving along stream-wise direction. The shock angles are obtained using image processing with edge detection filters. The incident shock angle, β_1 variation in time for shock-shock interaction is shown in Fig. 22.8. The figure also shows the rough error band along the data, which is around $\pm 1°$. It can be seen that the shock angle decreases for accelerating flow and increases for decelerating flow from Fig. 22.8. This trend is as expected. There is a small increase in the shock angle at the beginning of the nozzle movement in Fig. 22.8a, which may be nonphysical, and it needs to be explained in the future.

The Mach numbers corresponding to accelerated and decelerated flows are calculated by two methods: (1) from shock angles, and (2) from the ratio of settling chamber's stagnation pressure and the nozzle exit's static pressure. These plots for accelerating and decelerating flows are shown in Fig. 22.9. The error bars for pressure are wider than the error bar for the Mach number measurement from shock angles. The Mach numbers from the two methods agree very well with each other. This is true for both accelerating and decelerating cases. This suggests that the shock angles are as predicted by the steady state gas dynamics expressions. Further, from the schlieren images, it can be seen that there was no shock curvature or unsteady characteristics.

The wind tunnel was operated at a dM/dt of 1.67 for both acceleration and decelerating cases. The ratio of the time rate of change of the Mach number to the inverse of characteristic flow time scale is of the order of 2×10^{-5}. From Laguarda et al.

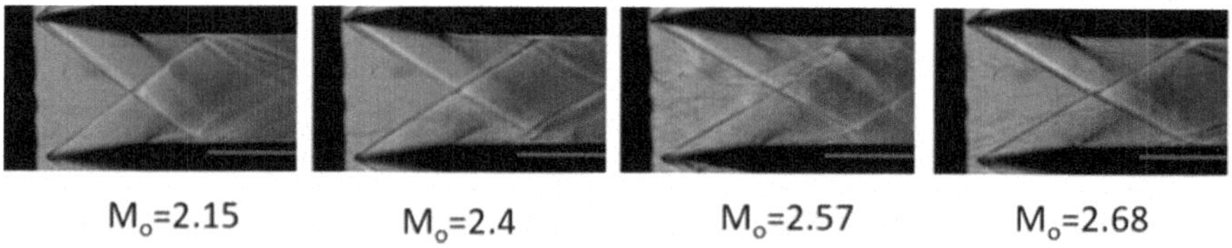

Fig. 22.6 Schlieren images of the accelerating flow at various time instants

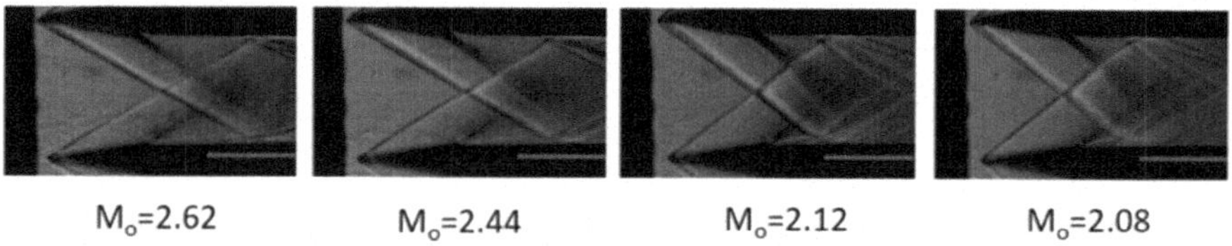

Fig. 22.7 Schlieren images of the decelerating flow at various time instants

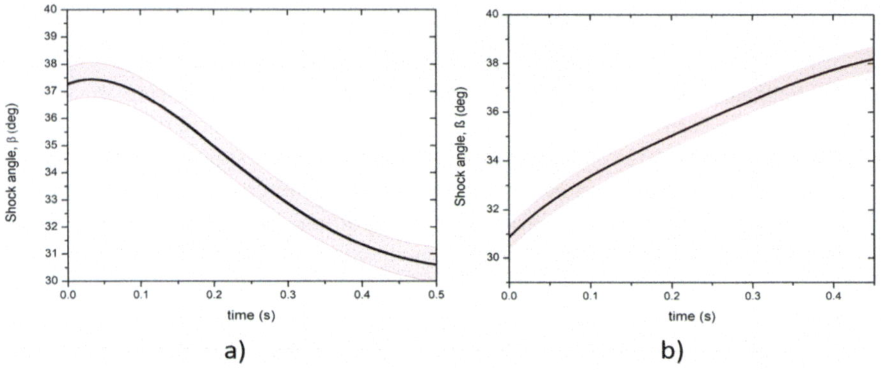

Fig. 22.8 Shock angle, β versus time for **a** acceleration **b** deceleration

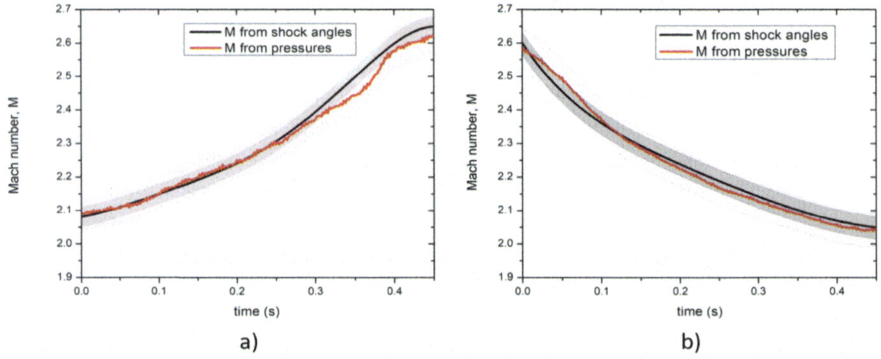

Fig. 22.9 Mach number, M versus time for **a** acceleration **b** deceleration

(2021), it can be said that the shock angle variation in the present study is quasi-steady [10], based on this. This is showing that the dM/dt of $1.67\,s^{-1}$ is still not sufficient to effect an unsteady flow physics that is significant in this flow field.

Table 22.1 gives the β_2 (refer Fig. 22.1) values from experimental measurements from images, at specific times in the accelerating run. Since the M_o was calculated using β_1 measurement, we are not comparing it with the theoretical values. The post-interaction shock angles β_2 and its theoretical value based on steady gas dynamics are presented.

It can be seen from the table that at every time instance presented, the data matches closely (within the error band) with the steady gas dynamics expressions based values of β_2. Suggesting that even the dM/dt of the $1.67\,s^{-1}$ was not enough to push this flow field into unsteady flow field. Similar results were observed even for the decelerating case as seen from Table 22.2.

The interaction of expansion fan with shock wave was attempted in this work, and a sample image (shown Fig. 22.10) is presented here in Fig. 22.10. It can be seen that there is not much deflection of the shock wave due to the expansion fan, and

Table 22.1 Comparison of shock angles after interaction of shocks, for four instances from the accelerating run

M_O	β_1 (deg)	$\beta_{2,theoretical}$ (deg)	$\beta_{2,experimental}$ (deg)
2.15	36.5	44.6	44.0 ± 3
2.4	33	39.3	41 ± 3
2.57	31.0	36.6	39.3 ± 3
2.68	29.9	35.0	36.0 ± 3

Table 22.2 Comparison of shock angles after interaction of shocks, for four instances from the decelerating run

M_O	β_1 (deg)	$\beta_{2,theoretical}$ (deg)	$\beta_{2,experimental}$ (deg)
2.62	30.6	36.0	36.5 ± 3
2.44	32.5	38.6	39.7 ± 3
2.12	37	45.6	48 ± 3
2.08	37.8	46.7	48.9 ± 3

Fig. 22.10 Schlieren image of the shock-expansion interaction model

thus it was very difficult to measure any change in shock angle beyond the error limits. Thus the interaction of shock with expansion fan was stopped. There needs to be more careful design of the wedges to effect a significant deflection of the shock wave.

22.4 Conclusion

Interaction of two oblique shocks generated by two wedges of 10° deflection was studied in a Variable Mach number tunnel. The shock angles from the wedges were used to calculate the instantaneous Mach number, based on steady gas dynamics expressions, as the tunnel Mach number was changing in time. The Mach number thus measured was then compared with the pressure ratio based exit plane Mach number. The values were matching very well, suggesting that the flow was almost behaving like a steady state flow. This was the case for both accelerating and decelerating cases. Suggesting that even the dM/dt of the $1.67\,s^{-1}$ was not enough to push this flow field into unsteady flow field. Thus the major conclusion from this work is that we need dM/dt higher than 1.67 to have unsteady gas dynamics effects. Experiments with expansion fan and oblique shock interactions, did not give sufficient deflections to be able to measure, and thus this experiment needs to be modified in the future.

References

1. G. Ben-Dor, M. Ivanov, E.I. Vasilev, T. Elperin, Hysteresis processes in the regular reflection ↔ Mach reflection transition in steady flows. Progress Aerospace Sci. **38**(4–5), 347–387 (2002)
2. H. Li, G. Ben-Dor, Oblique-shock/expansion-fan interaction-analytical solution. AIAA J. **34**(2), 418–421 (1996)
3. M.S. Ivanov, G. Ben-Dor, T. Elperin, A.N. Kudryavtsev, D.V. Khotyanovsky, Flow-Mach-number-variation-induced hysteresis in steady shock wave reflections. AIAA J. **39**(5), 972–974 (2001)
4. A. Durand, B. Chanetz, R. Benay, A. Chpoun, Investigation of shock waves interference and associated hysteresis effect at variable-Mach-number upstream flow. Shock Waves **12**(6), 469–477 (2003)
5. Y. Yao, S.G. Li, Z.N. Wu, Shock reflection in the presence of an upstream expansion wave and a downstream shock wave. J. Fluid Mech. **735**, 61 (2013)
6. L. Nel, B. Skews, K. Naidoo, Schlieren techniques for the visualization of an expansion fan/shock wave interaction. J. Visual. **18**(3), 469–479 (2015)
7. R. Goyal, A. Sameen, T. Jayachandran, G. Rajesh, Dynamic effects in transition from regular to Mach reflection in steady supersonic flows. Phys. Rev. E **104**(5), 055101 (2021)
8. L. Laguarda, J. Santiago Patterson, F.F.J. Schrijer, B.W. van Oudheusden, S. Hickel, Experimental investigation of shock-shock interactions with variable inflow Mach number. Shock Waves **31**(5), 457–468 (2021)
9. V. Vijayakrishnan, P. Kishore, M. TM, Investigation of variable mach number wind tunnel with symmetric sliding block nozzles, in AIAA SCITECH 2022 Forum, (2022) (p. 0716)
10. L. Laguarda, S. Hickel, F.F.J. Schrijer, B.W. Van Oudheusden. Dynamics of unsteady asymmetric shock interactions. J. Fluid Mech. 888 (2020)

Chapter 23
Shock Wave Characteristics of Twin-Supersonic Jet Interaction in a Confined Duct

Spandan Maikap and R. Arun Kumar

Abstract This study investigates the evolution of shock structures during the interaction of two rectangular parallel underexpanded jets in a confined duct. A comparison of shock structures during the interaction of twin jets in confined as well as with open boundaries were also investigated. An experimental study has been conducted in which a primary straight duct is opened into a suddenly expanding section forming a backward facing step (BFS) on the top and bottom sides of the duct outlet. A secondary underexpanded jet is issued, parallel to the primary jet, from the top BFS wall. The two supersonic jet interacts and merges, forming a single jet, which sticks to the nearest duct wall due to the Coanda effect. Optical flow visualization using the time-resolved shadowgraph method has been performed to investigate the evolution of shock reflection characteristics. Two interaction shocks are formed when the jets interact, which terminates at the triple point of the Mach reflections formed in the primary and secondary jet. The reflected shock waves from the Mach reflection of primary and secondary jet interacts with each other forming an Edney Type-V kind of reflection.

23.1 Introduction

The interaction of twin-parallel jets is widely seen in various aerodynamic systems. For instance, supersonic parallel jet interaction is prevalent in multi-jet rocket propulsion systems [1]. The interaction of rectangular supersonic jets is found in linear-type aerospike engines, in which a rectangular multi-jet system is used for thrust production [2]. The jet impingement study for powered electronics utilizes multiple parallel jets for cooling enhancement [3]. Another field of study where parallel jet interaction is commonly found is ACHEON (Aerial Coanda High-Efficiency Orienting-jet Nozzle) nozzles, where the jet thrust vectoring is possible without using any moving part [4]. In the ACHEON nozzles, the controlled jet is diverted towards the nozzle

S. Maikap · R. Arun Kumar (✉)
Indian Institute of Technology Jodhpur, Jodhpur, India
e-mail: arunkr@iitj.ac.in

© The Author(s), under exclusive license to Springer Nature Singapore Pte Ltd. 2024
G. Rajesh et al. (eds.), *Advances in Shock Interactions*, Springer Proceedings in Physics 408, https://doi.org/10.1007/978-981-97-6099-2_23

walls by interacting the jet with a secondary jet and the merged jet sticks to the wall with the help of the Coanda effect. Cen et al. [5] investigated the effect of a curved Coanda surface on the thrust vectoring of a supersonic jet. They found that adding a backward step in the 'Coanda flap' introduced a control response; however, the overall deflection angle of the primary jet was reduced. In the case of an underexpanded jet, the flow fields of the jet are predominantly governed by its shock structures. Moreover, the interaction of two underexpanded jets produces complex shock structures which can exert thermal loading and induce backflows and flow separation in the system. Therefore, it is imperative to study the shock interactions developed during the interaction of two underexpanded jets.

The interaction of two parallel jets separated by a distance is analogous to a single jet imping sidewise on a solid wall, where the midplane of the two jets behaves like a symmetric wall [6]. The jets are separated from the midplane up to a certain streamwise distance, and then they merge together, which is analogous to the reattachment of a single jet on the wall. Therefore, the jet characteristics in a twin jet interaction are quite different from a single jet, and such involute behaviour of twin jets results in complex shock interactions in the case of supersonic jets [1]. Prior works have been done by Moustafa et al. [7, 8], where two high-speed axisymmetric jets interact in the open to atmosphere condition. The study showed that the two jets merge together and become a single jet in the downstream section. It was also reported that the nozzle pressure ratio (NPR) and the inter-jet spacing play a significant role in the flow field characteristics. Interaction of two overexpanded jets produced by two identical Laval nozzles also showed similar flow characteristics [8]. Wlezien [9] reported that in the case of twin supersonic jet interaction, for closely spaced jets, the coupling occurs at lower Mach number and decreases with higher Mach number. The opposite of this trend is observed in the case of widely spaced jets. Desikan et al. [10] investigated the shock structures involved in the starting characteristics of twin underexpanded circular jets using Schlieren flow visualization. They found that due to the interaction of the two jets, several unique shock waves such as the 'trailing shocks' and 'terminal shocks' started appearing, which are not observed in a single unexpanded jet. A detailed review of the twin supersonic jet interaction with rectangular nozzle geometry is given by [11].

It is thus seen from the literature that the twin supersonic jet interaction characteristics have been widely investigated in the past. However, most of these studies were limited to jet interaction in open space, and the jet interaction in confined ducts bounded by walls has not been investigated rigorously. In many practical applications, the supersonic jet interaction happens in confined passages, and it is expected that the wall effects can bring profound changes to the well-understood supersonic jet interaction characteristics in open space. It was also reported that the mechanism of shock transition (from Mach reflection to Regular reflection) in supersonic underexpanded jet in confined passage shows a significant difference compared to the shock transitions in underexpanded supersonic jet in open space [12]. The present study, thus, investigates the interaction of twin underexpanded rectangular parallel jets in a confined duct and its shock wave interaction characteristics.

23.2 Experimental Setup

The experiments for the current study were performed using the Blowdown jet facility at IIT Jodhpur. The experimental model used in the present work consists of a straight primary rectangular duct of cross section 10×40 mm^2, having an inlet connected to a settling chamber and an outlet opening to a suddenly expanding duct. The suddenly expanding duct has two backward facing steps on the top and bottom of the primary duct, as shown in Fig. 23.1. A rectangular opening for the secondary jet with a cross section area of 6×40 mm^2 is provided in the top and bottom BFS regions. Pressurized air for secondary jet injection is taken directly from a high-pressure tank, and the pressure of the air is controlled using diaphragm-based pressure regulators. Upstream to the secondary duct opening, a temporary stagnation chamber is provided where the circular intake of the pressurized air from the pipes is converted into a rectangular sonic jet. The blowdown jet facility provides the high-pressure air to the primary duct. The primary jet air pressure is controlled using an unloaded type feedback-based pressure regulator.

The current experimental study consists of flow visualization in the test section to investigate the shock interaction characteristics. Transient flow visualization was carried out by capturing the second gradient of density along the flow direction using the time-resolved shadowgraph imaging technique. A conventional Z-type optics system was used in the experiments. A halogen-based light source is used to illuminate the test area. The light is collimated by the first field mirror (concave mirror of 200 mm diameter and 1000 mm focal length). Images of the flow fields were captured at 1000 frames per second (fps) and 10 μs exposure time using a CHRONOS 2.1 HD high-speed camera.

(a)

(b)

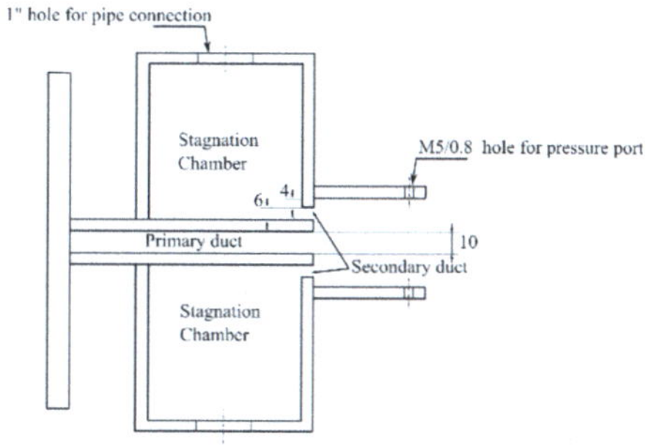

Fig. 23.1 **a** Experimental model, **b** schematics of the experimental setup

23.3 Results and Discussions

In this study, a choked jet is injected from the primary duct at a stagnation pressure of 3 bar. A choked secondary jet is injected at a stagnation pressure of 2.589 bar at the bottom side of the primary jet. The secondary jet is gradually opened using a solenoid valve, and the transient development of shock interactions has been investigated. Optical flow visualization using shadowgraph imaging has been used to investigate the shock interaction characteristics.

Figure 23.2a shows the shadowgraph image of the primary underexpanded jet issued in the duct without injecting any secondary jet. It should be noted that the initial primary jet is nearly symmetric in nature, and it is at the onset of an unstable state of 'pitch fork bifurcation'. Any perturbation given to the primary jet can shift the jet to one of its two stable states, that is the asymmetric expansion of the jet. Figure 23.2a shows that a Mach reflection structure (I) appears in the first shock cell of the primary jet. The schematics of the underexpanded jet is shown in Fig. 23.2b, describing its various compression and expansion waves. Figure 23.2c shows the shadowgraph image at the time instance where the secondary jet injection is initiated, and the mass flow rate is nearly 20% of the choked mass flow. As the secondary jet inflow occurs, several weak compression waves can be seen in the secondary duct region, which is being pushed out of the jet with an increase in secondary mass flow rate. These compression waves do not produce significant shock interaction characteristics with the primary jet. The fluid particles in the primary jet deflected by the expansion fans experience a collision with the secondary jet at the interaction point (T1) and change their direction to become parallel to the jet centerline of the primary jet. An oblique shock enables this deflection, referred to as 'primary jet interaction shock' (II). The primary jet interaction shock originates from the point of interaction (T1) and terminates close to the triple point produced by the the Mach reflection (I) structure in the primary jet. At the initial stages of the secondary jet injection, the 'primary jet interaction shock' interacts with the triple point located at the upperpart of the Mach reflection in the primary jet.

With further increase in the secondary jet mass flow, a 'secondary jet interaction shock' (III) starts developing, and a slip line (SL1) is created at the plane of the interaction of the jets, emerging from the point of interaction, as shown in Fig. 23.2e and f. Moreover, the primary jet interaction shock shifts its earlier position and terminates at the lower part of the triple point produced by the Mach reflection (I) in the primary jet. A reflected shock (V) can also be seen in the primary jet Mach reflection structure. Figure 23.2g shows the shadowgraph image when the jet interaction's reaches a steady state. A schematic representation of the shock structures has been described in Fig. 23.2h. The primary and secondary jet merge together, forming a single jet in the region. Due to the collision of the two jets, interaction shocks (II and III) emerge from the point of interaction (T1) in both the primary and secondary jets. The interaction shocks allow the fluid particles to deflect away from the interaction point and redirect their course of direction, making them parallel to the jet's centerline. The interaction shock waves in the primary (II) and secondary jet (III)

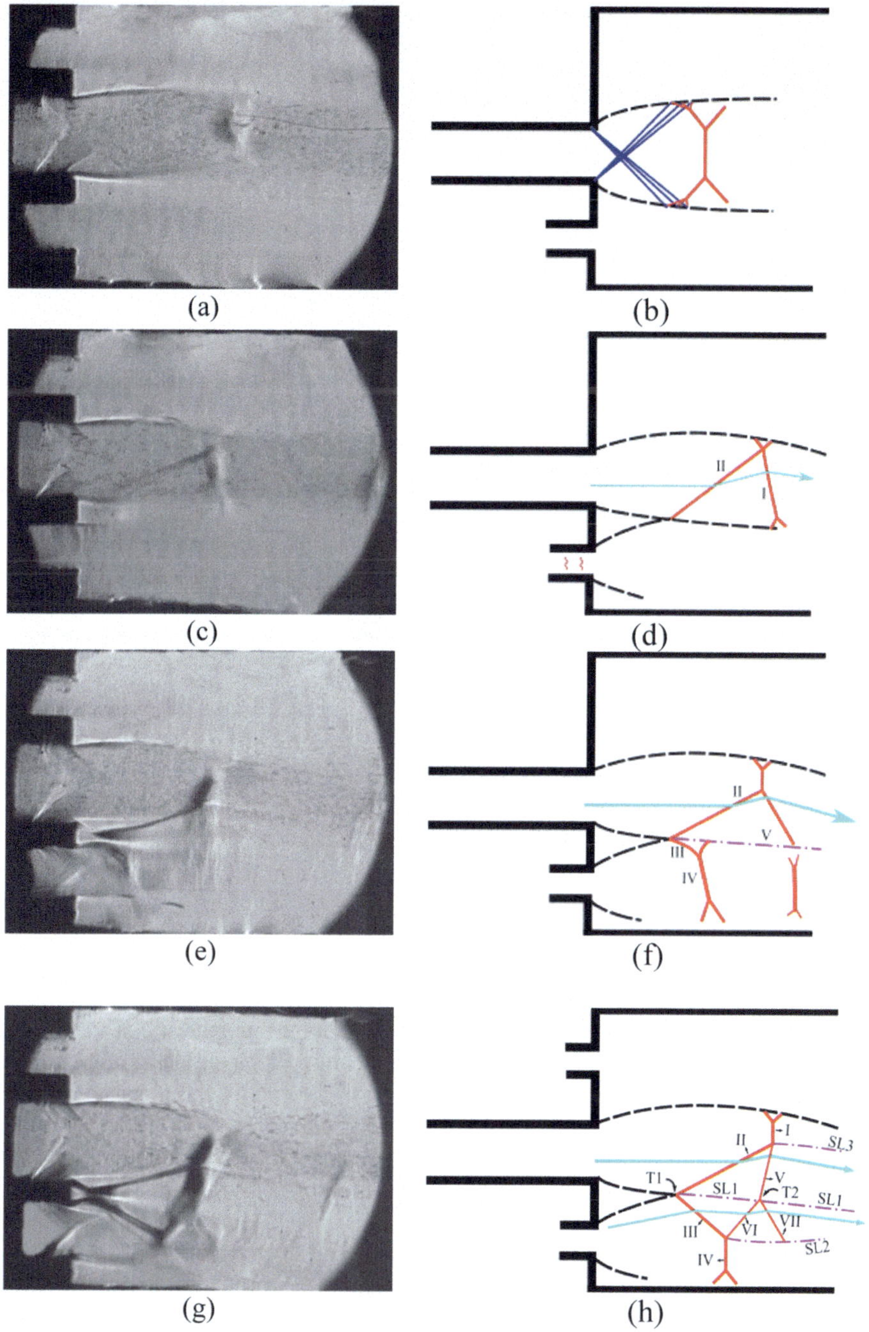

Fig. 23.2 Instantaneous shadowgraph images for the transient opening of the secondary jet. **a** 0% opening, **b** 25% valve opening time, **c** 75% of the valve opening time and **d** 100% of valve opening time

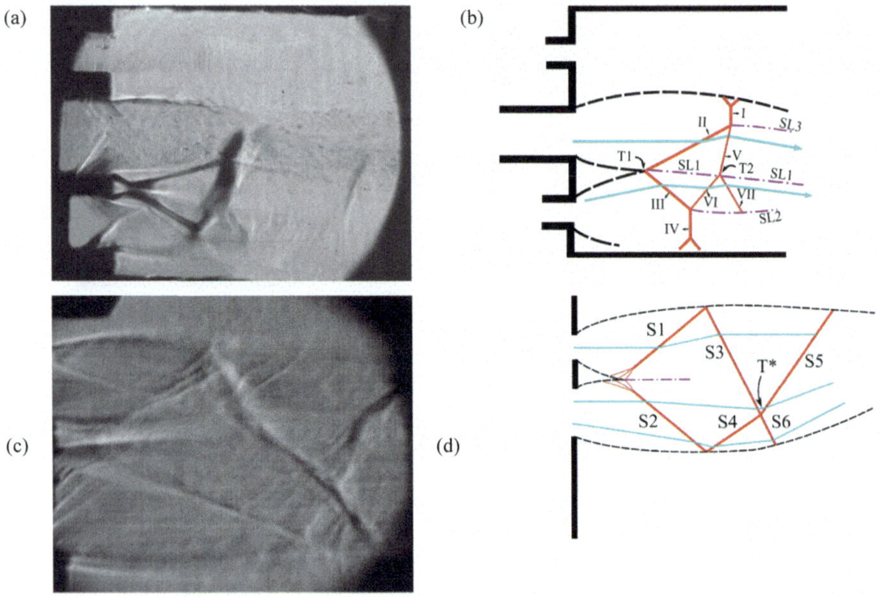

Fig. 23.3 **a** Shadowgraph image and **b** Schematic of the interaction of two underexpanded confined jets. **c** Shadowgraph image and **d** Schematic of two underexpanded free jet interactions

interacts with the respective Mach reflection shock structures (I and IV, respectively). In the primary jet the 'interaction shock' wave interacts with the lower part of the Mach reflection and in the secondary jet, the interaction shock interacts with the upper part of the Mach reflection. The interaction of the 'primary and secondary jet interaction shock' wave determines the location of the new triple point in upper side of the secondary jet and the triple point in the lower side of the primary jet. Reflected shock waves (V and VI) emantes from the new triple points formed by the interaction of the interaction shock with the Mach reflection in the primary and secondary jets. These reflected shock wave from the primary and secondary jet interacts with each other at a new triple point location. This type of shock interaction can be related to the type-III interaction from the Edney classification of shock interactions A slip stream can also be seen at the triple point location formed by the intraction of the reflected shock waves (V and VI) from the primary and secondary jets.

Figure 23.3 shows the comparison of cases in which (i) two underexpanded supersonic free jets and (ii) two underexpanded confined jets are made to interact. Figure 23.3a and b shows the shadowgraph images and schematic for the case when two underexpanded jets are made to interact. The description of the flow features has been discussed in detail in the previous paragraph. Figure 23.3c shows the shadowgraph images of the case when two free jets are interacting. The spacing of the two jets is kept the same as in the confined duct case. The supply pressure of both the jets is maintained at 4 bar. It is seen that the two jets merge together, interacting at an interaction point. The injection of the secondary jet does not deflect the primary jet in any direction. This is because the free jets do not experience the Coanda effect provided by the walls, which was seen in the confined jet interaction case. It can be seen that when the two jets interact, two oblique shocks (S1 and S2) are formed, which are similar to the interaction shocks formed in the confined duct. These oblique shocks (S1 and S2) originate near the interaction region and terminate at the jet boundary. The oblique shocks (S1 and S2) reflect from the jet boundary and form two more oblique shocks (S3 and S4). S3 and S4 interact at a point (T*), offset to the plane of interaction, towards the primary jet, and transmits S5 and S6, as shown in Fig. 23.3d. It should be noted that the reflected shocks (S3 and S4) interact as a type-II Edney shock interaction, unlike the type-III kind of interaction seen in confined jet interaction. Therefore, It can be observed that the shock structures seen in the case of supersonic jets interacting in free space and supersonic jets interacting in the confined passage are different in nature.

23.4 Conclusion

Two underexpanded rectangular parallel jets are made to interact in a confined passage, and the shock interactions formed due to the interactions are investigated. The primary and the secondary jet merge together and form a single jet. The presence of the walls casuses the merged jet to stick to the wall due to the Coanda effect. It is seen that several unique shock structures are formed when two rectangular parallel jets are made to interact in a confined passage. 'Primary and secondary jet interaction shocks' are observed, emerging at the point of interaction and terminating at the triple point of the respective mach reflections. Reflected shocks from the primary and secondary mach reflection meet together, forming a new triple point at a downstream location. Type-III kind of Edney's shock interaction is observed at this location.

The interaction of two underexpanded confined jets is compared with the case of two underexpanded free jets. In the case of free jet interaction, the two free jets merge, forming a single jet. However, the merged jet does not change its direction with the secondary jet's injection. The interaction of the secondary jet leads to the formation of interaction shock waves in the primary and secondary jet. The interaction shocks reflect back from the jet boundary as two oblique shocks. The interaction of the reflected oblique shocks creates a type-II Edney shock interaction at a point offset to the plane of the interaction of the two jets. Thus, the study inferred that the shock structure interaction formed due to the confined jet interactions is significantly different from the free supersonic jet interaction.

References

1. P.V. Raje, K. Sinha, Three-dimensional simulation of rocket nozzles with multi-jet interaction using shock-unsteadiness model, in *AIAA Aviation 2019 Forum* (2019), pp. 3322–3337

2. T. Niimi, H. Mori, K. Okabe, Y. Masai, M. Taniguchi, Analyses of flow field structures around linear-type aerospike nozzles using LIF and PSP, in *20th International Congress on Instrumentation in Aerospace Simulation Facilities, ICIASF'03* (IEEE, 2003), pp. 15–20
3. R. Leena, G. Syamkumar, M.J. Prakash, Experimental and numerical analyses of multiple jets impingement cooling for high-power electronics. IEEE Trans. Components, Packag. Manuf. Technol. **8**(2), 210–215 (2018)
4. M. Song, S. Park, Y. Lee, Application of backstep Coanda flap for supersonic coflowing fluidic thrust-vector control. AIAA J. **52**(10), 2355–2359 (2014)
5. Z. Cen, T. Smith, P. Stewart, J. Stewart, Integrated flight/thrust vectoring control for jet-powered unmanned aerial vehicles with ACHEON propulsion. Proc. Inst. Mech. Eng. Part G: J. Aerosp. Eng. **229**(6), 1057–1075 (2015)
6. A. Nasr, J.C.S. Lai, Comparison of flow characteristics in the near field of two parallel plane jets and an offset plane jet. Phys. Fluids **9**(10), 2919–2931 (1997)
7. G.H. Moustafa, Experimental investigation of high-speed twin jets. AIAA J. **32**(11), 2320–2322 (1994)
8. G.H. Moustafa, Interaction of axisymmetric supersonic twin jets. AIAA J. **33**(5), 871–875 (1995)
9. R.W. Wlezien, Nozzle geometry effects on supersonic jet interaction. AIAA J. **27**(10), 1361–1367 (1989)
10. S.L.N. Desikan, B. Murugan, K. Srinivasan, S. Sajan, Twin jet interaction and reverse flow. J. Spacecraft Rockets **52**(6), 1577–1585 (2015)
11. G. Raman, P. Panickar, K. Chelliah, Aeroacoustics of twin supersonic jets: a review. Int. J. Aeroacoustics **11**(7–8), 957–984 (2012)
12. R. Arun Kumar, G. Rajesh, Shock transformation and hysteresis in underexpanded confined jets. J. Fluid Mech. **823**, 538–561 (2017)

The manufacturer's authorised representative in the EU is Springer Nature Customer Service Centre GmbH, Europaplatz 3, 69115 Heidelberg, Germany. If you have any concerns regarding our products, please contact ProductSafety@springernature.com

Printed and bound by CPI Group (UK) Ltd, Croydon, CR0 4YY
01/10/2026
02225964-0005